T0259825

LESSONS ON SOIL

Snowdon from Llyn Llydaw, showing the first stage in soil formation.

LESSONS ON SOIL

BY

SIR E. J. RUSSELL, D.SC., F.R.S.

President of the British Association, Late Director of the Rothamsted Experimental Station, Harpenden

CAMBRIDGE
AT THE UNIVERSITY PRESS
1950

CAMBRIDGE
UNIVERSITY PRESS

University Printing House, Cambridge CB2 8BS, United Kingdom

Published in the United States of America by Cambridge University Press, New York

Cambridge University Press is part of the University of Cambridge.

It furthers the University's mission by disseminating knowledge in the pursuit of education, learning and research at the highest international levels of excellence.

www.cambridge.org
Information on this title: www.cambridge.org/9781107684782

First edition 1911
Reprinted 1912, 1920, 1922, 1926, 1930
Second edition 1950
First published 1950
First paperback edition 2014

A catalogue record for this publication is available from the British Library

ISBN 978-1-107-68478-2 Paperback

CONTENTS

LIST OF ILLUSTRATIONS

LIST OF ILLUSTRATIONS

LIST OF ILLUSTRATIONS

PREFACE TO THE NEW EDITION

This little book has had a very long life because it deals with something that never loses its interest and importance—the soil on which we depend for our food, and which carries the flowers and grass and trees that give us so much pleasure. It was written for the scholars at the village school of Wye in Kent, to whom I used to give a lesson each week, at the invitation of the Headmaster, the late Mr W. J. Ashby; it was carefully revised at St George's School, Harpenden, when the Rev. Cecil Grant gave me a class of bright, critical Third Form boys and girls who knew that the lessons were going to be published—they had therefore been well tested before they were offered to other teachers.

Since those days wars and social changes have greatly altered our countryside and have heavily reduced the area of agricultural land left to us. Even during the short lifetime of this book we have added 7 million to our population in England and Wales, and lost 3 million acres of agricultural land, and the losses still go on, more rapidly than ever. Our financial position makes it more and more imperative that we should import less food and either produce more ourselves, or go without. It is now, therefore, supremely important to every one of us that the fullest and best use should be made of our remaining soil, and I hope this book will help in giving a wider understanding of some of its wonderful properties and possibilities.

Many changes have been made in the text to bring it into line with the modern conditions in the countryside, and I wish to thank Dr E. W. Russell for much valuable criticism and help in making the revision. Finally, it is a pleasure to thank Mr Martin Fayers for much friendly help right from the old days at Wye.

E. JOHN RUSSELL

CAMPSFIELD WOOD, WOODSTOCK
November 1949

INTRODUCTION

The following pages contain the substance of lessons given at the village school at Wye to the Fourth, Fifth, Sixth and Seventh Standards (mixed) and at St George's School, Harpenden, to the Third Form. There is, however, an important difference between the actual lessons as given and the book. The lessons had reference to the soils round about the village, and dealt mainly with local phenomena, general conclusions being only sparingly drawn; while in the book it has been necessary to throw the course into a more generalized form. The teacher in using the book will have to reverse the process, he must find local illustrations and make liberal use of them during the course; it is hoped that the information given will help him over any difficulties he may experience.

This necessity for finding local illustrations constitutes one of the fundamental differences between the study of Nature and other subjects of the school curriculum. The text-book in some of the others may be necessary and sufficient; in dealing with Nature it is at most only subsidiary, serving simply as a guide to the thing that is to be studied; unless the thing itself be before the class it is no better than a guide to a cathedral would be without the cathedral. And just as the guide is successful only when he directs the attention of the stranger to the important features of the place, and fails directly he becomes garrulous and distracts attention, so a Nature Study book succeeds only in as far as it helps in the study of the actual thing, and fails if it is used passively and is substituted for an active study. No description or illustration can take the place of direct observation; the simplest thing in Nature is infinitely more wonderful than our best word pictures can ever paint it.

The author recommends the teacher to look through the chapter before it has to be taken in class and then to make a few expeditions in search of local illustrations. It is not strictly necessary that the chapters should be taken in the order given. The local phenomena must be dealt with as they arise and as weather permits, or the opportunity may pass not to return again during the course. In almost any lane, field, or garden a sufficient number of illustrations may be obtained; if a stream and a hill are accessible the material is practically complete, especially if the scholars can be induced to pursue their studies during their summer-holiday rambles. Of course, this entails a good deal of work for the teacher, but the results are worth it. Young people enjoy experimental and observation lessons in which they take an active part and are not merely passive learners. The value of such lessons in developing their latent powers and in stimulating them to seek for knowledge in the great book of Nature is a sufficient recompense to the enthusiastic teacher for the extra trouble involved.

It is not desirable to work through a chapter in one lesson. Scholars unaccustomed to make experiments or to see experiments done, will probably require three or four lessons for getting through each of the first few chapters, and two or three lessons for each of the others.

The pot experiments of chapters VI, VII and VIII should be started as early in the course as possible. Twenty flower pots are wanted for the set; they should be of the same size, about 8 inches being a convenient diameter, and should be kept together in a warm place. Three are filled with sand, seven with subsoil, and the remaining ten with surface soil. Three of the subsoil pots are uncropped, two being stored moist and one dry. Four pots of the surface soil are uncropped and moist, a fifth and sixth are uncropped and dry, one of these contains earthworms (p. 44). Four glazed pots, e.g. large jam or marmalade jars, are also wanted (p. 56). Mustard, buckwheat or rye make good crops, but many others will do. Leguminous crops, however, show certain abnormal characters, while turnips and cabbages are apt to fail; none of these should be used. It is highly desirable that the pots should be duplicated.

The plots also should be started in the school garden as early as convenient. Eight are required for the set; their treatment is described in chapter IX. Plots 2 yards square suffice.

A supply of sand, of clay, and of lime will be wanted, but it is not necessary to have fresh material for each lesson. The sand may be obtained from a builder, a sand pit, the seashore or from a dealer in chemical apparatus. The clay may be obtained from a brick yard; it gives most satisfactory results after it has been ground ready for brick making. Modelling clay is equally satisfactory. A supply of rain water is desirable.

For a class of twelve pupils working in pairs at the experiments the following apparatus is wanted for the whole course:

Six tripods and bunsen burners or spirit lamps [2].
Twelve pipe-clay triangles [4].
Twelve crucibles or tin lids [3].
Sixteen gas jars [4].
Twelve beakers 250 c.c. capacity [4].
Two beakers 500 c.c.
Two beakers 100 c.c.
Six egg-cups [2].
Twelve funnels [3].
Six funnel stands [1].
Six perforated glass disks [3].
Two tubulated bottles 500 c.c., four corks to fit.
Cork borers.
4 lb. assorted glass tubing.
Pestle and mortar.

Twelve Erlenmeyer flasks 50 c.c. [3].
Six saucers.
Twelve flat-bottomed flasks 100 c.c., six fitted with india-rubber stoppers bored with one hole [3], and six with ordinary corks [3].
Box as in Fig. 13.
Six glass tubes ½ inch diameter, 18 inches long [2].
Six straight lamp chimneys [3].
Six test-tubes, corks to fit.
Three Factory thermometers with stems 6 inches long.
Soil sampler (p. 78).
Balance and weights.
Two retort stands with rings and clamp.
Soil acidity indicator (p. 54).

The figures given in square brackets are the quantities that suffice when the teacher alone does the experiments, it not being convenient for the scholars to do much. They are the figures given at the head of each chapter.

CHAPTER I

WHAT IS THE SOIL MADE OF?

APPARATUS REQUIRED.* Soil and subsoil from a hole dug in the garden. Clay. Two tripods and bunsen burners or spirit lamps. Two crucibles or tin lids and pipe-clay triangles. Four jars or gas cylinders. Two beakers.

If we talk to a farmer or a gardener about soils he will say that there are several kinds of soil—clay soils, gravel soils, peat soils, chalk soils, and so on—and we may discover this for ourselves if we make some rambles in the country and take careful notice of the ground about us, particularly if we can leave the road and walk on the footpaths across the fields. When we find the ground very hard in dry weather and very sticky in wet weather we may be sure we are on a clay soil, and may expect to find brick yards or tile works somewhere near, where the clay is used. If the soil is loose, drying quickly after rain, and if it can be scattered about by the hand like sand on the seashore, we know we are on a sandy soil and can look for pits where builder's sand is dug. But it may very likely happen that the soil is something in between, and that neither sand pits nor clay pits can be found; if we ask what sort of soil this is we are told it is a loam. A gravel soil will be known at once by its gravel pits, and a chalk soil by the white chalk quarries and old lime kilns, while a peat soil is black, sometimes marshy and nearly always spongy to tread on.

We want to learn something of the soil round about us, and we will begin by digging a hole about 3 feet deep to see what we can discover. At Harpenden this is what the scholars saw: the top 8 inches of soil was dark in colour and easy to dig; the soil below was reddish brown in colour and very hard to dig; one changed into the other so

* The numbers represent the requirement when the experiments are done by the teacher alone or by one group of scholars.

quickly that it was easy to see where the top soil ended and the bottom soil began; no further change could, however, be seen below the 8-inch line. A drawing was made to show these things, and is given in Fig. 1. You may find something quite different: sand, chalk, or solid rock may occur below the soil, but you should enter whatever you see into your note-books and make a drawing, like Fig. 1, to be kept for future use. Before filling in the hole, some of the dark-coloured top soil, and some of the lighter coloured soil lying below (which is called the SUBSOIL), should be taken for further examination; the two samples should be kept separate and not mixed.

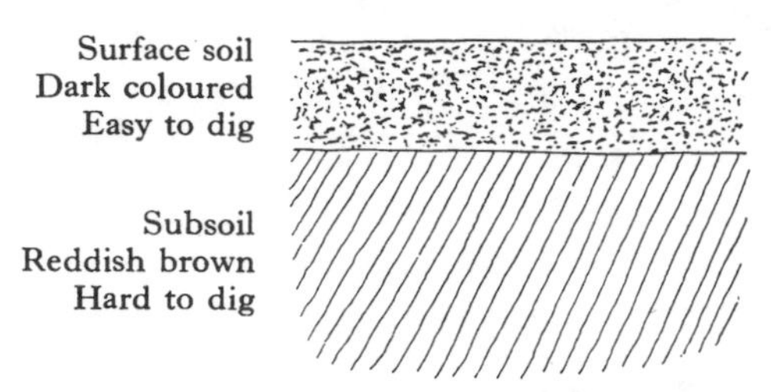

Fig. 1. Soil and subsoil in St George's school garden.

First look carefully at the top soil and rub some of it between your fingers. We found that our sample was wet and therefore contained water; it was very sticky like clay and therefore, presumably, contained clay; there were a few stones and some grit present and also some tiny pieces of dead plants—roots, stems or leaves, but some so decayed that we could not quite tell what they were. A few pieces of a soft white stone were found that marked on the blackboard like chalk.* Lastly, there were a few fragments of coal and cinders but, as these were not a real part of the soil, we supposed they had got in by accident. The subsoil was also wet and even more sticky than the top soil; it contained stones and grit, but seemed almost free from plant remains and from the white chalky fragments.

* Later on we found other soils that did not contain these white fragments.

A few experiments will show how much of some of these things are present. The amount of water may be discovered by weighing out 10 grams of soil, leaving it to dry in a warm place near the fire or in the sun, and then weighing it again. In one experiment the results were:

Weight of top soil before drying	10 grams	= 100 decigrams
Weight of top soil after drying	8·3 grams	= 83 decigrams
Water lost	1·7 grams	= 17 decigrams

A column 100 millimetres long was drawn to represent the 100 decigrams of soil, and a mark was drawn 17 millimetres up to show the amount of water (see Fig. 2).

Weight of bottom soil before drying	10 grams	= 100 decigrams
Weight of bottom soil after drying	8·7 grams	= 87 decigrams
Water lost	1·3 grams	= 13 decigrams

Another column should be drawn for the subsoil. On drying the soil it becomes lighter in colour and loses its stickiness, but it has not permanently changed, because as soon as water is added it comes back to what it was before.

The dried lumps of soil are now to be broken up finely with a piece of wood, but nothing must be lost. It is easy to see shrivelled pieces of plant, but not easy to pick them out; the simplest plan is to burn them away. The soil must be carefully tipped on to a tin lid, or into a crucible, heated over a flame and stirred with a long clean nail or bradawl. First of all it chars, then there is a little sparkling, but not much, finally the soil turns red and does not change any further no matter how much it is heated. The shade of red will at once be recognized as brick red or terra cotta; indeed

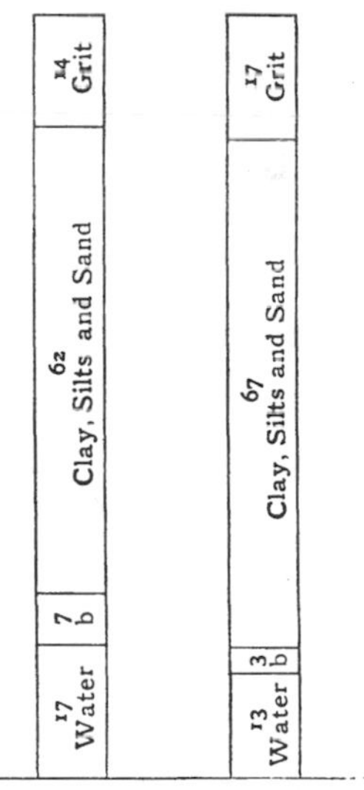

Surface soil Subsoil

b stands for the part that burns away

Fig. 2. Columns showing what 100 parts of surface soil and subsoil were made of.

'terra cotta' means 'baked earth'. When the soil is cold it should be examined again; it has become very hard and the plant remains have either disappeared or have changed to ash and crumble away directly they are touched. On weighing, a further loss is discovered, which was in our experiment:

Weight of dry top soil before burning	83 decigrams
Weight of dry top soil after burning	76 decigrams
The part that burnt away weighed	7 decigrams
Weight of dry subsoil before burning	87 decigrams
Weight of dry subsoil after burning	84 decigrams
The part that burnt away weighed	3 decigrams

These results are entered on the column in Fig. 2.

The surface soil is seen to contain more material that will burn away than the subsoil does. When the burnt soil is moistened it does not become dark and sticky like it did before, it has completely changed and cannot be made into soil again. It is more like brick dust than soil.

For further experiments we shall want a fresh portion of the original soil.

On a wet afternoon something was noticed that enabled us to get a little further with our studies. The rain water ran down a sloping piece of ground in a tiny channel it had made; the streamlet was very muddy, and at first it was thought that all the soil was washed away. But we soon saw that the channel was lined with grit, some of which was moving slowly down and some not at all. GRIT can therefore be separated from the rest of the soil by water.

This separation can be shown very well by the following experiment. Rub 10 grams of finely powdered soil with a little water (rain water is better than tap water), keep it still for 10 seconds,* then carefully pour the muddy liquid into a large glass jar. Add more

* Instead of using a watch you can adopt a device often used by photographers: say at ordinary conversational speed '1 little second, 2 little seconds...' up to '10 little seconds' and with a little practice this takes approximately 10 seconds.

water to the rest of the soil, shake, pause for 10 seconds, and again pour the liquid into the jar; go on doing this till the jar is full. Then get some more jars and still keep on till the liquid is no longer muddy but nearly clear. The part of the soil that remains behind and will not float over into the jars is at once seen to be made up of small stones, grit, and sand. Set the jars aside and look at them after a day or so. The liquid remains muddy for some time, but then it clears and a thick black sediment gathers at the bottom. If now you very carefully pour the liquid off you can collect the sediments: they are soft and sticky, and can be moulded into patterns like clay. In order to see if they really contain clay we must do the experiment again, but use pure clay from a brick yard, or modelling clay, instead of soil. The muddy liquid is obtained as before, it takes a long time to settle, but in the end it gives a sediment so much like that from the soil, except in colour, that we shall be safe in saying that the sediments in the jars contain the CLAY from the soil. And thus we have been able to separate the sticky part of the soil—the clay—from the gritty or sandy part which is not at all sticky. We may even be able to find out something more. If we leave the soil sediment and the clay sediment on separate tin lids to dry, and then examine them carefully we may find that the soil sediment is really a little more gritty than the clay. Although it contains the clay, it also contains something else.

When the experiment is made very carefully in a proper way this material can be separated from the clay and divided into fine sand, and a still finer material called SILT. But these divisions are not sharp: the coarse sand shades off into the fine sand, the fine sand shades off into the silt, and the silt into the clay. Between the clay and the sand there are great differences, though it would be impossible to say exactly where clay ends and silt begins.

If there is enough grit it should be weighed: we obtained 14 decigrams of grit from 10 grams of our top soil and 17 decigrams from

10 grams of bottom soil. We cannot separate the clay from the silt, but when this is done in careful experiments it is usually found that the subsoil contains more clay than the top soil. We should of course expect this because we have found that the subsoil is more sticky than the top soil. These results are put into the columns as

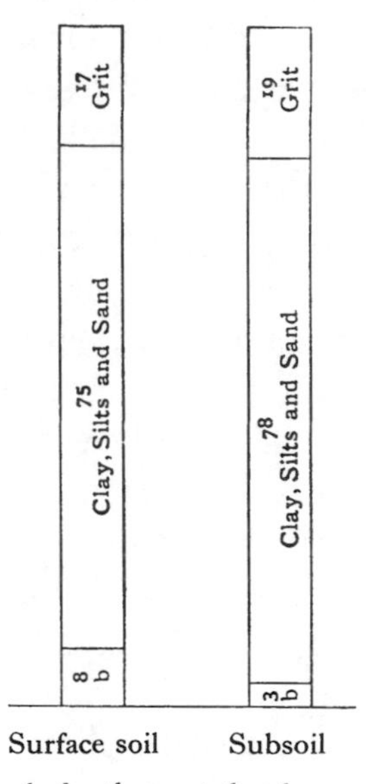

Fig. 3. Columns showing what 100 parts of dried surface soil and subsoil were made of.

before so that we can now see at once how much of our soil is water, how much can burn away, how much is grit, and how much is clay and other things.

What would have happened if the sample had been dug out during wetter or drier weather? The quantity of water would have been different, but in other respects the soil would have remained the same. It is therefore best to avoid the changes in the amount of water by working always with 10 grams of *dried* soil. The results we obtained were:

	Top soil	Subsoil
Weight of dry soil before burning	100	100 decigrams
Weight of dry soil after burning	92	97 decigrams
The part that burned away weighed	8	3 decigrams
Weight of grit from 10 grams of dried soil	17	19 decigrams

The columns are given in Fig. 3.

SUMMARY. The experiments made so far have taught us these facts:

1. Soil contains water, grit or sand, silt, clay, and a part that burns away; some soils also contain white chalky specks.

2. The top layer of soil to a depth of about 8 inches is different from the soil lying below, which is called the subsoil. It is less sticky, easier to dig, and darker in colour. It contains more of the material that burns away, but less clay than the subsoil.

3. When soil is dried it is not sticky but hard or crumbly; as soon as it is moistened it changes back to what it was before. But when soil is burnt it completely alters and can no longer be changed back again.

CHAPTER II

MORE ABOUT THE CLAY

Apparatus Required. Clay about 6 lb.; a little dried, powdered clay; sand, about 6 lb. Two glass jars or cylinders. One beaker. One egg-cup. Two funnels and stands. Two perforated glass or tin disks. Two glass tubes. Two tubulated bottles fitted with corks. Some seeds. Two small jars about 2 inches × 1 inch. Bricks. The apparatus in Fig. 9. Pestle and mortar.

We have seen in the previous chapter that clay will float in water and only slowly settles down. Is this because clay is lighter than water? Probably not, because a lump of clay seems very heavy. Further, if we put a small ball of clay into water it at once sinks to the bottom. Only when we rub the clay between our fingers or work it with a stick—in other words, when we break the ball into very tiny pieces—can we get it to float again. We therefore conclude that the clay floated in our jars (p. 4) for so long, not because it was lighter than water, but because the pieces were so small.

Clay is exceedingly useful because it can be moulded. Dig up some clay, if there is any in your garden, or procure some from a brick works. You can mould it into any shape you like, and the purer the clay the better it acts. Enormous quantities of clay are used for making bricks. Make some model bricks about an inch long and half an inch in width and depth, also make a small basin of about the same size, then set them aside for a week in a warm, dry place. They still keep their shape; even if a crack has appeared the pieces stick together and do not crumble to a powder.

If you now measure with a ruler any of the bricks that have not cracked, you will find that they have shrunk a little and are no longer quite an inch long. This fact is well known to brickmakers; the moulds in which they make the bricks are larger than the brick is wanted to be. But what would happen if instead of a piece of clay 1 inch long you had a whole field of clay? Would that shrink also, and, if so, what would the field look like? We can answer this question in two ways; we may make a model of a field and let it dry, and we can pay a visit to a clay meadow after some hot, dry weather in summer. The model can be made by kneading clay up under water and then rolling it out on some cardboard or wood as if it were a piece of pastry. Cut it into a square and draw lines on the cardboard right at the edges of the clay. Then put it into a dry warm

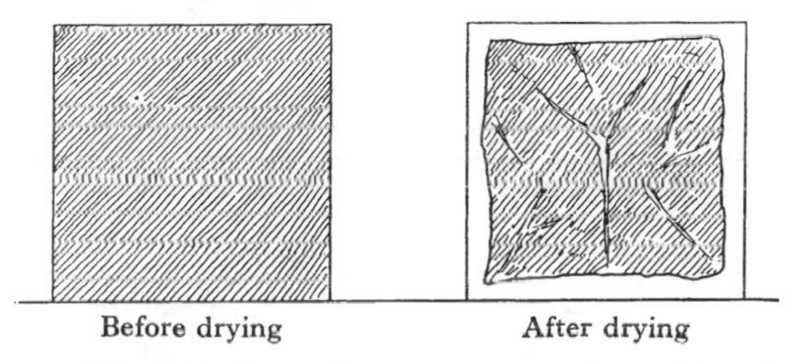

Fig. 4. Clay was plastered over a square piece of board and completely covered it. After drying for a week the clay had shrunk and cracked.

place and leave for some days. Fig. 4 is a picture of such a model after a week's drying. The clay has shrunk away from the marks, but it has also shrunk all over and has cracked. If you get an opportunity of walking over a clay field during a dry summer, you will find similar but much larger cracks, some of which may be 2 or 3 inches wide, or even more. Sometimes the cracking is so bad that the roots of plants or of trees are torn by it, and even buildings, in some instances, have suffered through their foundations shrinking away. We can now understand why some of our model bricks cracked. The cracks were caused by the shrinkage just as happens with our model field. As soon as the clay becomes wet it swells again. A very pretty experiment can be made to show this. Fill a glass tube or an egg-cup with dry powdered clay, scrape the surface level with a ruler, and then stand it in a glass jar full of rain water so that the whole is completely covered. After a short time the clay begins to swell and forces its way out of the egg-cup as shown in Fig. 5, falling over the side and making quite a little shower. In exactly the same way the ground swells after heavy rain and rises a little, then it falls again and cracks when it becomes dry. Darwin records some careful measurements in a book called *Earthworms and Vegetable Mould*—'a large flat stone laid on the surface of a field sank 3·33 millimetres* whilst the weather was dry between 9 May and 13 June, and rose 1·91 millimetres between 7 and 19 September of the same year, much rain having fallen during the latter part of this time. During frosts and thaws the movements were twice as great.'

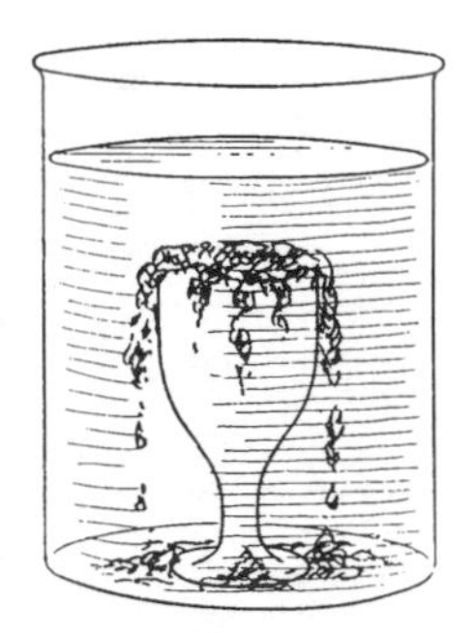

Fig. 5. Clay swelling up when placed in water and overflowing from the egg-cup into which it was put.

You must have found out by now how very slippery clay becomes as soon as it is wet enough. It is not easy to walk over a clay field in wet weather, and if the clay forms part of the slope of a hill it may be so slippery that it becomes dangerous. Sometimes after very heavy rains soil resting on clay on the side of a hill has begun to slide downwards and moves some distance before it stops. Fortunately these LAND SLIPS, as they are called, are not common in England, but they do occur. Fig. 6 shows one in Devonshire, and another is described by Gilbert White in *The Natural History of Selborne*.

Another thing that you will have noticed is that anything made of clay holds water. A simple way of testing this is to put a round piece of tin perforated with holes into a funnel, press some clay on to it and on to the sides of the funnel (Fig. 7), and then pour on rain water. The water does not run through. Pools of water may lie like

* A little more than one-eighth of an inch.

this on a clay field for a very long time in winter before they disappear, as you will know very well if you live in a clay district. So when a lake or a reservoir is being made it sometimes happens that the sides are lined with clay to keep the water in.

Fig. 6. Landslip at Bindon, near Axminster, Devon.

If water cannot get through can air? This is very easily discovered: plug a glass tube with dry powdered clay and see if you can draw or blow air through. You can. But now use wet clay: air can no longer get through. Clay can be used like putty to stop up holes or cracks, and so long as it keeps moist neither air nor water can pass through. Take two bottles like those in Fig. 8, stop up the bottom tubes, and fill with water. Then put a funnel through each cork and fit the cork in tightly, covering with clay if there is any sign of a leak. Put a perforated tin disk into each funnel, cover one well with clay and the other with sand. Open the bottom tubes. No water runs out from the first bottle because no air can leak in through the clay, but it runs out very quickly from the second because the sand lets air

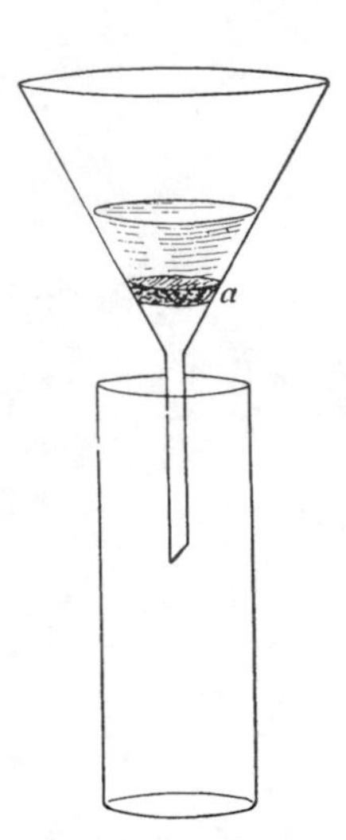

Fig. 7. A thin layer of clay *a* entirely prevents the water running through.

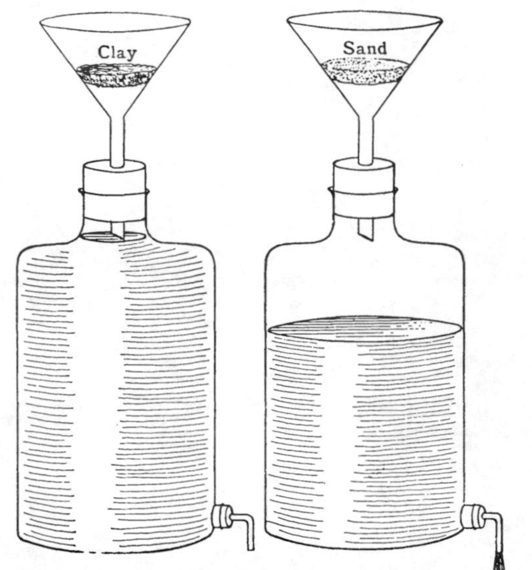

Fig. 8. Sand allows air to pass through it. and so water runs out of the bottle. Clay does not let air pass, and the water is therefore kept in, even though the tube is open.

through. These properties of clay and sand are very important for plants. Sow some small seeds in a little jar full of clay kept moist to prevent it cracking, and at the same time sow a few in some moist sand. The seeds soon germinate in the sand but not in the clay. It is known that seeds will not germinate unless they have air and water and are warm enough. They had water in both jars, and they were in both cases warm, but they got no air through the clay and therefore could not sprout. Pure clay would not be good for plants to grow in. Air came through the sand, however, and gave the seeds all they wanted for germination.

This also explains something else that you may have noticed. If you tried baking one of your model bricks in the fire you probably found that the brick exploded and shattered to pieces: the water still left in the brick changed to steam when it was heated, but the steam could not escape through the clay, and so it burst the clay. In a brickworks the heat is very gradually applied and the steam only slowly forms, so that it has time to leak away, then when it has all gone the brick can be heated strongly. You should try this with one of your model bricks; leave it in a hot place near the stove or on the radiator for a week or more and then see if you can bake it without mishap.

Let us now compare a piece of clay with a brick. The differences are so great that you would hardly think the brick could have been made from clay. The brick is neither soft nor sticky, and it has not the smooth surface of a piece of clay, but is full of little holes or pores, which look as if they were formed in letting the steam out. A brick lets air through; some air gets into our houses through the bricks even when the windows are shut. Water will get through bricks more easily than it does through clay. After heavy rain you can often find that water has soaked through a brick wall and made the wallpaper quite damp. A pretty experiment can be made with the piece of apparatus shown in Fig. 9: bore in a brick a hole about an inch deep and a quarter of an inch wide, put into the hole the piece of bent glass tubing, and fix it in with some clay or putty, then pour some water blackened with ink in the tube, marking its position with a label.

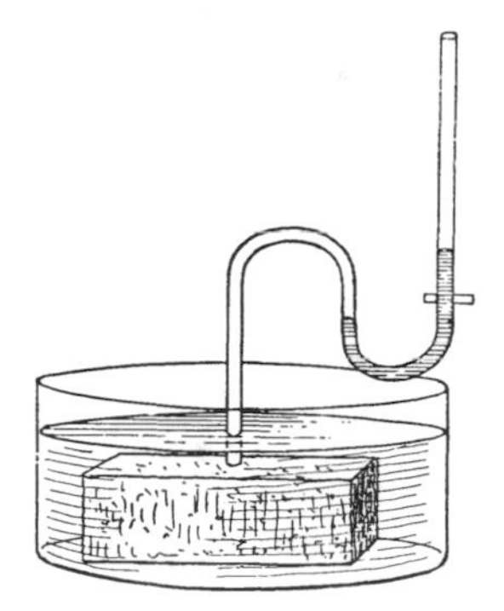

Fig. 9. A brick standing in water. The air in the brick is driven inwards by the water and forces the liquid up the tube in order to escape.

Stand the brick in a vessel so full of water that the brick is entirely covered. Water soaks into the brick and presses the air out: the air tries to escape through the tube and forces up the black liquid.

One more experiment may be tried. Can a brick be changed back into clay? Grind up the brick and it forms a gritty powder. Moisten it, work it with your fingers how you please, but it still remains a gritty powder and never takes on the greasy, sticky feeling of pure clay. Indeed no one has ever succeeded in making clay out of bricks. All these experiments show that clay is completely altered when it is burnt. We also found that soil is completely altered by burning, and if you look back at your notes you will see that the changes are very much alike, so much so that we can safely put down some of the changes in the burnt soil—the red colour, the hard grittiness, and the absence of stickiness—to the clay. Let us now examine a piece of dry, but unburnt, clay. It is very hard and does not crumble, it is neither sticky nor slippery. Directly, however, we add some water it changes back to what it was before. Drying therefore alters clay only for the time being, whilst baking changes it permanently.

CHAPTER III

WHAT LIME DOES TO CLAY

APPARATUS REQUIRED. Clay, about 6 lb. Some of the clay from chapter II may, if necessary, be used over again. Lime, about ½lb. Two funnels, stands and disks. Two glass jars. Lime water.*

If you are in a clay district in autumn or early winter you will find some of the fields dotted with white heaps of chalk or lime, and you will be told that these things 'improve' the soil. We will make a few experiments to find out what lime does to clay. Put some clay on to a perforated tin disk in a funnel just as you did on p. 10, press it down so that no water can pass through. Then sprinkle on to the clay some powdered lime and add rain water. Soon the water begins to leak through, though it could not do so before; the addition of the lime, therefore, has altered the clay. If you added lime to a garden or a field on which water lay about for a long time in winter you would expect the water to drain away, especially if you made drains or cut some trenches along which the water could pass. There are large areas in England where this has been done with very great advantage.

The muddy liquid obtained by shaking clay with water clears quickly if a little lime is stirred in. Fill two jars *A* and *B* (Fig. 10) with rain-water, rub clay into each and stir up so as to make a muddy liquid, then add some lime water to *B* and stir well. Leave for a short time. Clots quickly appear in *B*, then sink, leaving the liquid clear, but *A* remains cloudy for a long time. But why should the liquid clear? We decided in our earlier experiments that the clay floated in the water because it was in very tiny pieces; when we took

* Lime water is made by shaking up lime and water. It should be kept in a well-corked bottle.

a larger lump the clay sank. The lime has for some reason or other, which we do not understand, made the small clay particles stick together to form the large clots, and these can no longer float, but sink. If we look at the limed clay in our funnel experiment we shall see that the same change has gone on there; the clay has become rather loose and fluffy, and can therefore no longer hold water back.

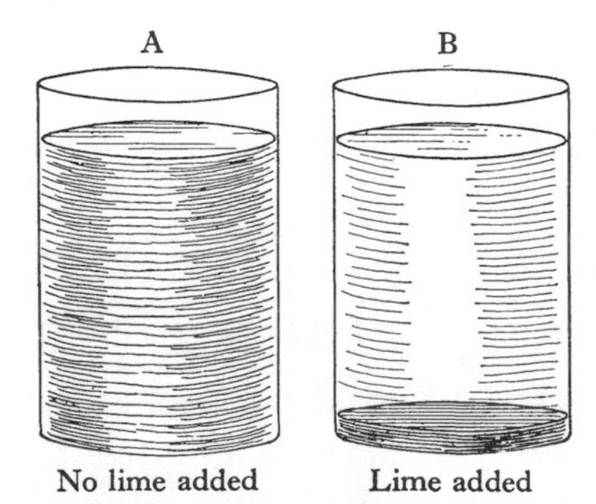

Fig. 10. Addition of lime to turbid clay water now makes the clay settle and leaves the water quite clear.

Lime also makes clay less sticky. Knead up one piece of clay with rain water alone and another piece with rain water and about one-twentieth its weight of lime. The limed clay breaks easily and works quite differently from the pure clay.

SUMMARY. This, then, is what we have learnt about clay. Clay is made up of very tiny pieces, so small that they float in water. They stick together when they are wetted and then pressed, and they remain together; a piece of clay moulded into any pattern will keep its shape even after it is dried and baked. Clay is therefore made into bricks, earthenware, pottery, etc., whilst white clay, which is found in some places, is made into china. Wet clay shrinks and cracks as it dries; these cracks can easily be seen in the fields during dry weather. This shrinkage may interfere with the foundations of houses and other buildings, causing them to settle. Dry clay is different from wet clay; it is hard, not sticky and not slippery, but it at once becomes like ordinary clay when water is added. After baking, however, clay permanently alters and cannot again be changed back to what it was before. Clay will not let water pass through; a clay field is therefore nearly always wet in winter and spring. Nor can air pass through until the clay dries or cracks.

WHAT LIME DOES TO CLAY

Lime has a remarkable action on clay. It makes the little, tiny pieces stick together to form feathery clots which sink in water; lime therefore causes muddy clay water to become clear. The clots cannot hold water back, and hence limed clay allows water to pass through. Limed clay is also less sticky than pure clay. A clay field or garden is improved by adding lime because the soil does not remain wet so long as it did before; it is also less sticky and therefore more easily cultivated.

CHAPTER IV

SOME EXPERIMENTS WITH THE SAND

APPARATUS REQUIRED. Sand, about 6 lb.; clay, about 6 lb. Funnel, stand and disk. Two glass jars. One box with glass front shown in Fig. 13 filled with clay and sand as indicated. Quarry chalk (about 5 lb.). A beaker and an egg-cup.

If there is a sand pit near you, or a field of sandy soil, you should get a supply for these experiments; if not, some builder's sand can be used. When the sand is dry, you will see that the grains are large and hard. Further, they are all separate and do not stick together; if you make a hole in a heap of the sand, the sides fall in; there is nothing solid about it. When the sand is wet, it sticks better and can be made into a good many things; at the seaside you can make a really fine castle with wet sand. But as soon as the sand dries, it again becomes loose and begins to fall to pieces.

Strong winds will blow these fragments of dry sand about and pile them up into the sand hills or DUNES common in many seaside districts (Fig. 11). BLOWING SANDS can also be found in inland districts; in the northern part of Surrey, in parts of Norfolk and many other places are fields where so much of the soil is blown away by strong winds that the crops may suffer injury. In Central Asia sand storms do very much harm and have in the course of years buried entire cities. Fig. 12 shows the Penhale sands in Cornwall gradually covering up some meadows and ruining them. The ancient city of Kenfig near what is now Port Talbot was buried in this way; in 1949 a Norman building was discovered lying under 70 feet of sand.

Sand particles, being large, do not float in water. If we shake up sand in water the sand sinks, leaving the water entirely clear. So running water does not carry sand with it unless it is running very quickly: the sand lies at the bottom.

Sand holds water much less firmly than clay does. Pour some water on to sand placed on the tin disk in a funnel (Fig. 8); much of it runs through at once. We should therefore expect a sandy field or a sandy road to dry up very quickly after rain and not to remain

Fig. 11. Sand dunes, Penhale sands, Cornwall.

wet like a clay field. So much is this the case that people prefer to live on a sandy rather than on a clay soil. The most desirable residential districts round London, Hampstead on the north, and the stretch running from Haslemere on the south-west to Maidstone on the south-east, and other favoured regions, are all on the sand.

At the foot of a hill formed of sand you often find a SPRING, especially if clay or solid rock lies below. It is easy to make a model that will show why the spring forms at this particular place. Fill the lower part of the box shown in Fig. 13 with wet clay, smoothing it out so that it touches all three sides and the glass front; then on top of the clay put enough sand to fill the box. Bore four holes in the side as shown in the picture, one at the bottom, one at the top, one just above the junction of the sand and clay, the fourth half way

Fig. 12. Sand from Penhale sand dunes blowing on to and covering up meadows.

up the sand, and fix in glass tubes with clay or putty. Pour water on to the sand out of a watering can fitted with the rose so as to imitate the rain. At first nothing seems to happen, but if you look closely you will notice that the water soaks through and does not lie on the surface; it runs right down to the clay; then it comes out at the tube there (*c* in the picture). None goes through the clay, nor does enough stay in the sand to flow out through either the top or the second tube; of the four tubes only one is discharging any water. The discharge does not stop when the supply of water stops. The rain need only fall at intervals, but the water can flow all the time.

The experiment should now be tried with some chalk from a quarry; it gives the same results and shows that chalk, like sand, allows water readily to pass.

Just the same thing happens out of doors in a sandy or chalky country; the rain water soaks through the sand or chalk until it comes to clay or solid rock that it cannot pass, then it stops. If it can

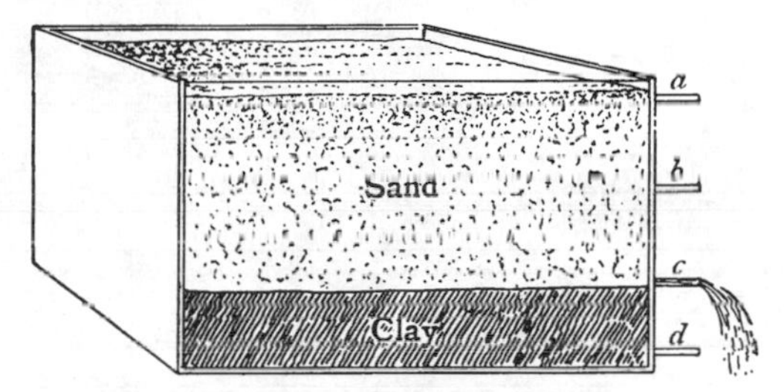

Fig. 13. Model spring. A box with glass front contains a layer of clay and one of sand. Water that falls on the sand runs right down to the clay but can get no farther, and therefore flows out through the tube *c* at the junction of the clay and the sand. The same result is obtained when chalk takes the place of sand.

find a way out it does so and makes a spring, or sometimes a whole line of springs or wet ground. Rushes, which flourish in such wet places, will often be found growing along this line, and may, indeed, in summer time be all you can see, the water having drained away. But after much rain the line again becomes very wet. Fig. 14 shows the foot of a chalk hill near Harpenden, where a spring breaks out just under the bush at the right-hand side of the gate. In Fig. 15 the bush itself is seen, with the little pool of water made by the spring. Here the water flows gently, but elsewhere it sometimes happens, as in Fig. 16, that the spring breaks out with great force.

Now stop up the glass tubes so that the water cannot get out. Soon the sand becomes flooded and is no better than clay would be.

A second model will show this very well. Make a large saucer of clay and fill with sand: pour water on. The water stays in the sand, because it cannot pass through the clay. A sandy field saturated like this will therefore not be dry, but wet, and will not make a good position for a house. We must therefore distinguish the two cases

Fig. 14. Foot of a chalk hill at Harpenden where a spring breaks out.

illustrated in Fig. 17. *A* shows sand on a hill, dry because the water runs through until it comes to clay or rock, when it stops and breaks out as a spring, a tiny stream, or pond; this is a good building site and you may expect to find large houses there. *B* shows the sand in a basin of clay, where the water cannot get away: here the cellars and downstairs rooms are liable to be wet, and in a village the wells give impure water. Matters could be improved if a way out were cut for the water, but this may not always be practicable.

It often happens that villages are situated at the junction of sand and clay, or chalk and clay, because the springs furnish a good water supply. Look at the map of the chalk Down country from Kent to Wiltshire and you will find a string of villages at the foot of the Downs where the springs break out.

Fig. 15. The little pool and the spring.

On the other hand, large tracts of clay which remain wet and sticky during a good part of the year are not very attractive to live in, and even near London they were the last to be populated: many people can remember Hither Green in the south-east and the clay districts of the north-west being built up, but the sands and gravels of Highgate, Chiswick, Brentford and other places had long been occupied. Elsewhere, villages on the clay do not grow quickly unless there is much manufacturing or mining; the parishes are large, and before the days of motoring the roads were not good, while in the old days they used to be very bad indeed. Macaulay tells us that at the end of the seventeenth century in some parts of Kent and Sussex 'none but the strongest horses could, in winter, get through the bog, in which at every step they sank deep. The markets were

Fig. 16. Water spouting up from a bore hole (artesian water), Old Cateriag Quarry, Dunbar.

often inaccessible during several months....The wheeled carriages were, in this district, generally pulled by oxen. When Prince George of Denmark visited the stately mansion of Petworth in wet weather, he was six hours in going nine miles; and it was necessary that a body of sturdy hinds should be on each side of his coach, in order to prop it. Of the carriages which conveyed his retinue several were upset and injured. A letter from one of the party has been preserved

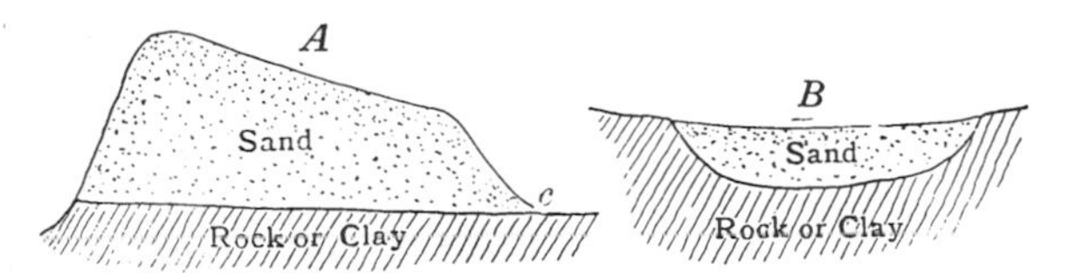

Fig. 17. Two positions of sand. *A* is dry because the water can drain away and break out as a spring at *c*. *B* is wet because the water cannot drain away.

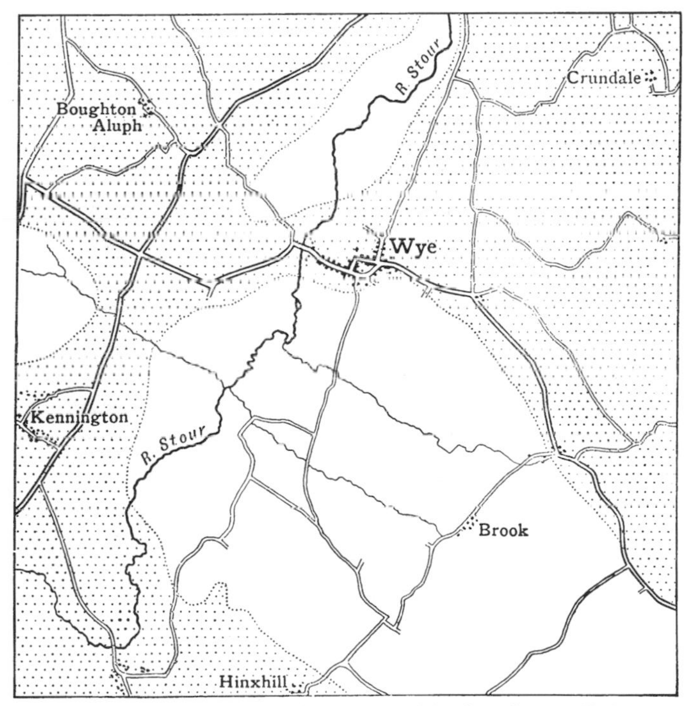

Fig. 18. The roads round Wye. As far as possible they keep off the clay (the plain part of the map) and keep on the chalk or the sand (the dotted part of the map).

in which the unfortunate courier complains that, during fourteen hours, he never once alighted, except when his coach was overturned or stuck fast in the mud.' The Romans knew how to make roads anywhere, and so they made them run in a straight line between the two places they wished to connect, but the art was lost in later years, and the country roads made in England since their time usually had to follow the sand or the chalk, avoiding the clay as much as possible. These roads we still use. Fig. 18 shows the roads round Wye; you should in your rambles study your own roads and see what soil they are on. Nowadays, however, roads can be made on any kind of soil.

There are several other ways in which sand differs from clay. It does not shrink on drying nor does it swell on wetting, and you will find nothing happens when you try with sand the experiment with the model field (p. 9) or the egg-cup (p. 10). More accurate experiments, however, show that dry sand shrinks somewhat on wetting.

CHAPTER V

THE PART THAT BURNS AWAY

APPARATUS REQUIRED. Leaf mould. Mould from a tree. Peat. About 1 lb. soil from a wood, a well-manured garden and a field; also some subsoil. Six crucibles or tin lids. Six tripods, pipe-clay triangles, and bunsen burners or spirit lamps. One beaker and an egg-cup.

In the autumn leaves fall off the trees and form a thick layer in the woods. They do not last very long; if they did they would in time almost bury the wood. You can, in the spring or early summer find out what has happened to them if you go into a wood or carefully search under a big hedge in a lane where the leaves were not swept away. Here and there you come across skeleton leaves where only the veins are left, all the rest having disappeared. But generally where the leaves have kept moist they have changed to a dark brown mass which still shows some of the structure of a leaf. This is called LEAF MOULD. The top layer of soil in the wood is soft, dark in colour, and is evidently leaf mould mixed with sand or soil.

Leaf mould is highly prized by gardeners, indeed gardeners will often make a big heap of leaves in autumn and let them 'rot down' and change into mould. If you can in autumn collect enough leaves to make a heap you should do so and leave it somewhere where the rain can fall on it, but cover it with a few small branches of trees to prevent the wind blowing the leaves away. The heap shrinks a great deal during the first few months, and in the end it gives a supply of mould that will be very useful if you want to grow any plants in pots.

Some of the little hollows in the bank under a hedge, especially on chalky soils, are filled with leaf mould which has sometimes changed to a black powder not looking at all like leaves.

You can also find mould in holes in decayed trees; here it has formed from the wood of the tree.

It appears, then, that dead leaves, etc., slowly change into a black or brown substance, shrinking very much as they do so. For this reason they do not go on piling up year after year till finally they fill the wood; instead they decay or 'rot down' to form leaf mould: the big pile of the autumn has changed by the next summer to a thin layer which mixes with the soil.

We want now to see what happens on a common or a piece of waste ground that is not cultivated. Grass and wild plants grow up in the summer and die during winter; their stems and roots are not taken away, but clearly they do not remain where they are, because next year new plants grow up. We may suppose that the dead roots and stems decay like the leaves did, and change to a brown or black mould. It looks as if we are right, because on digging a hole or examining the side of a freshly cut ditch we shall find that the top layer of soil, just so far as the living roots go, is darker in colour than the layer below.

We must, however, try and get some more proof, and to do this we must study some of our specimens a little more closely. Take some leaf mould, some black mould from a hollow in the bank, some from a tree, soils from a wood, a well-manured garden, a field and some subsoil. All except the subsoil have a dark colour, but the wood and garden soils are probably darker than the field soil. Now weigh out 2 grams of each of these and heat in a dish as you did the soil on p. 3; notice that all except the subsoil go black and then begin to smoulder, but the moulds smoulder more than the soils. Then weigh again and calculate how much has burnt away in each case. The following are some results that have been obtained at Harpenden.

The mould nearly all burns away and its dark colour entirely goes, so also does the dark colour of the soil.

	Colour before burning	Amount of smouldering	Percentage loss on burning	Colour of residue
Leaf mould	Dark brown	Much	78·3	Light grey
Mould from dead tree	Black	Much	60·6	Light grey
Soil from wood	Dark brown	Less	43·4*	White
Soil from garden	Almost black	Less	10·1	Red
Soil from field	Brownish	Still less	5·4	Red
Subsoil	Red	None	2·0	Red

Our supposition explains why, for many soils (though not for all), the less the blackness, the less the loss on burning. If the brown or black combustible part is really mould formed by the decay of plant roots, etc., then we should expect that as the percentage of mould in the soil increased, so its blackness would increase and its loss on burning would become greater. This actually happens.

But this does not mean that *all* the loss on burning a soil is due to plant material, for it is not: nor is it true that all black soils are rich in mould; some contain black minerals such as shale, which is quite different from mould.

This, then, is our idea. We suppose that the plants that have lived in past years have decayed to form a black material like leaf mould which stops in the soil, giving it a darkish colour. The more mould there is, the darker the colour of the soil. We know that along with this decay there is a great deal of shrinkage. The black material that is formed from the plant only extends as far into the soil as the plant roots go, so that there is a sharp change in colour about 6 inches below the surface (see also p. 2). Like the plant the black material formed from it all burns away when the soil is heated sufficiently.

Thus we can explain all the facts we have observed, and in what seems a very likely way. This does not show that our supposition is correct, but only that it is probable. When you come to study science subjects you will find such suppositions, or HYPOTHESES as

* The top 2 inches of soil only were collected here and there were many leaves, twigs, etc. mixed in. Soils from different woods vary considerably. If the sample is taken to a greater depth the loss on burning is much less, and may be only 4 or 5 per cent.

they are called, are frequently used so long as they are found to be helpful. In our present case we could only get absolute proof that the black combustible part of the soil really arose from the decay of plants by watching the process of soil formation. We shall turn later to this subject.

The black material is known as HUMUS. Farmers and gardeners like a black soil containing a good deal of humus because they find it very rich, and we shall see later on why this is so. Vast areas of such soils occurring in Manitoba, in Russia, and in Hungary are used for wheat growing.

In some places the black material forms a layer which may be several feet thick: it is then called PEAT. Near Wye there was a wet level area covered with about 2 feet of peat; it was fibrous and had evidently been formed from plants. Ditches had been cut in it to let the water escape and these showed that clay lay underneath; a sample was taken and tested as shown in Fig. 7: it held up the water completely. The peat bed was partly surrounded by higher ground from which water could drain but it could not easily escape so it stopped there: indeed the bed might easily have been a shallow lake. It was clear that any plants growing there would decay in very wet conditions, entirely different from those in the much drier soils we have just been studying.

The Fens round the Wash used to be like this, but they were properly drained, and they then proved to be extremely fertile (p. 98). In many places patches of low-lying peat have been drained and have then been very good for potatoes, vegetables, and many other crops.

But there is another kind of peat found in level stretches of high-lying country with high rainfall; heather and cotton-grass are the common plants. Here also the water cannot escape, but it is rain water and not land drainage water, and also the vegetation is different from that in the low-lying areas. So the peat is different, and it is not so good for growing plants as the low-lying peat, though it can be greatly improved by the addition of lime, as we shall see later on.

Look at a piece of peat and notice how very fibrous it is, quite unlike leaf mould. When it is dry, peat easily burns and is much used as fuel in parts of Scotland, Wales and Ireland. It is cut in blocks

Fig. 19. Peat cut for fuel and stacked for drying. Solway Moss, Cumberland.

during the spring, left to dry in heaps during summer, and then carried away in autumn. Fig. 19 shows a peat bog where cutting has been done. Peat does not easily catch light and the fires are generally kept burning all night; there is no great flame such as you get with a coal fire, but still there is quite a nice heat.

Peat has a remarkable power of absorbing water. Fill an egg-cup with peat, packing it as tightly as you possibly can, and then put it under water and leave for some days. The peat becomes very wet and swells considerably, overflowing the cup just like the clay did on p. 10. After long and heavy rains peat in bogs swells up so much that it may become dangerous; if the bog is on the side of a hill, the peat may overflow and run down the hill like a river, carrying everything before it. Such overflows sometimes occur in Ireland and they used to occur in the north of England; you can read about one on Pendle Hill in Harrison Ainsworth's *Lancashire Witches*. But they do not take place when ditches are cut in the bog so that the water can flow away instead of soaking in; this has been done in England.

This great power of absorbing water and other liquids, so terrible when it leads to overflows, enables peat to be put to various uses. A good deal of it is sold as PEAT-MOSS, for use in stables; and gardeners add it to composts for potting because it holds the moisture so well.

In the ditches of a peat bog red slimy masses can often be found. They look just like rusty iron, and in fact they do contain a good deal of iron, but there are also a number of tiny little living things present. The stones and grit just under the peat are usually white, all the red material from them having been washed out by the water which has soaked through the peat. Then at the ditch these tiny living things take up the red material because it is useful to them. Peat or 'moorland' water can also dissolve lead from lead pipes and may therefore be dangerous for drinking purposes unless it is specially purified. When you study chemistry you will be able to show that both peat itself and moorland waters are 'acid' while good mould is not. That is why peat is not good for cultivated plants (see also p. 68).

Other things besides peat are formed when plants decay under water. If you stir up the bottom of a stagnant pond with a stick, bubbles of gas rise to the surface and will burn if a lighted match is put to them. This gas is called marsh gas. Very unpleasant and unwholesome gases are also formed.

CHAPTER VI

THE PLANT FOOD IN THE SOIL

APPARATUS REQUIRED. The pot experiments (p. xvi).

It is a rare sight in England to see land in a natural uncultivated state devoid of vegetation. The hills are covered with grasses and bushes, the moors with ling and heather, commons with grass, bracken and gorse, a garden tends to become smothered in weeds, and even a gravel path will not long remain free from grass. It is clear that soil is well suited for the growth of plants. We will make a few experiments to see what we can find out about this property of soil.

We have seen that a good deal of the soil is sand or grit, and we shall want to know whether this, like soil, can support plant life. We have also found that the subsoil is unlike the top soil in several ways, and so we shall want to see how it behaves towards plants. Fill a pot with soil taken from the top 9 inches of an arable field or untrenched part of the garden; another with subsoil taken from the lower depth, 9 to 18 inches, and a third with clean builder's sand or washed sea-sand. Sow with rye or mustard, and thin out when the seeds are up. Keep the pots together and equally well supplied with water; the plants then have as good a chance of growth in one pot as in any other.

Figs. 20 and 21 are photographs of sets of plants grown in this way; the weights in grams were:

	Green weight		After drying	
	Rye	Mustard	Rye	Mustard
Plants grown in top soil (pot 3)	14·5	17·7	5·6	2·6
Plants grown in subsoil (pot 4)	2·9	5·1	1·6	1·1
Plants grown in sand (pot 5)	2·0	4·6	0·8	1·0

Fig. 20. Rye growing in surface soil (pot 3), subsoil (pot 4) and sand (pot 5).

The plants in the soil remained green and made steady growth. Those in the sand never showed any signs of getting on, their leaves turned yellow and fell off; in spite of the care they received, and the water, warmth and air given them, they looked starved, and that, in fact, is what they really were. Nor did those in the subsoil fare much

Fig. 21. Mustard growing in surface soil (pot 3), subsoil (pot 4), and sand (pot 5).

better. The experiment shows that the top soil gives the plant something that it wants for growth and that it cannot get either from sand or from the subsoil; this something we will call 'plant food'.

Further proof is easily obtained. At a clay or gravel pit little or no vegetation is to be seen on the sloping sides or on the level at the bottom, although the surface soil is carrying plants that shed innumerable seeds. A heap of subsoil thrown up from a newly made well, or the excavations of a house, lies bare for a long time. The practical man has long since discovered these facts. A gardener is most particular to keep the top soil on the top, and not to bury it, when he is trenching. In levelling a piece of ground for a cricket pitch or tennis court, it is not enough to lift the turf and make a level surface; the work has to be done so that at every point there is sufficient depth of top soil in which the grass roots may grow.

HOW MUCH PLANT FOOD IS THERE IN THE TOP SOIL? To answer this question we must compare soil that has been cropped, with soil that has been kept FALLOW, i.e. moist but uncropped. Tip out some of the soil that has been cropped with rye, and examine it. Remove the rye roots, then replace the soil in the pot and sow with mustard; sow also a fallow pot with mustard. Keep both pots properly watered. The soil that has carried a crop is soon seen to be much the poorer of the two. Fig. 22 shows the plants, while their weights in grams were:

	Green weight	After drying
Mustard growing in soil previously cropped with rye (pot 1)	17·8	3·3
Mustard growing in soil previously uncropped (pot 2)	62·3	8·6

The rye has taken most of the plant food that was in pot 1, leaving very little for the second crop. Our soil therefore contained only a little plant food, not more, in fact, than will properly feed one crop. But yet it did not seem to have altered in any way, even in weight, in consequence of the plant food being taken out. In our experiment the soil was dried and weighed before and after the mustard was grown; the results were:

	Pot 2 lb.	Pot 2 oz.	Pot 2*a* lb.	Pot 2*a* oz.
Weight of dried soil before the experiment	6	6	6	7
Weight of dried soil after the experiment	6	6	6	6
Difference	0	0	0	1

The experiment is not good enough to tell us exactly how much plant food was present at the beginning. But we can say that the amount of plant food in the soil is too small to be detected by such weighing as we can do.

Here is an account of a similar experiment made 300 years ago by van Helmont in Brussels, and it is interesting because it is one of the first scientific experiments on plant growth:

'I took an earthen vessel in which I put 200 lb. of soil dried in an oven, then I moistened with rain water and pressed hard into it

Fig. 22. Mustard growing in soil previously cropped with rye (pot 1), and in soil previously uncropped (pot 2).

a shoot of willow weighing 5 lb. After exactly five years the tree that had grown up weighed 169 lb. and about 3 oz. But the vessel had never received anything but rain water or distilled water to moisten the soil (when this was necessary), and it remained full of soil which was still tightly packed, and lest any dust from outside should have got into the soil it was covered with a sheet of iron coated with tin but perforated with many holes. I did not take the weight of the leaves that fell in the autumn. In the end I dried the soil once more, and got the same 200 lb. that I started with, less about 2 oz. Therefore the 164 lb. of wood, bark and root arose from the water alone.' The experiment is wonderfully good and shows how very little plant food there is in the soil. The conclusion is not quite right, however, although it was for many years accepted as proof of an ancient belief that all things arose from water. It is now known that the last sentence should read, 'Therefore the 164 lb. of wood, bark and root arose chiefly from the water *and air*, but a small part came from the soil also.'

But to return to our experiment with pots 1 and 2. They had been kept moist before the mustard was sown. Did this moisture have any effect on the soil? Take two of the pots that have been kept dry and uncropped, and two that have been kept moist and uncropped, also one of dry uncropped subsoil and one of moist uncropped subsoil. Sow rye or mustard in each pot and keep them all equally supplied with water.

It is soon evident that the top soil is richer in plant food than the subsoil, and the soil stored moist is rather richer than that stored dry, although the difference here is less marked. In an experiment in which the soils were put up early in July and sown at the end of September the weights of crops in grams obtained were:

	Pot no.	Green weight	After drying
Plants grown in top soil stored in moist condition	10	16·9	2·6
	11	18·9	2·8
Plants grown in top soil stored in dry condition	8	12·1	1·8
	9	14·4	1·9
Plants grown in subsoil stored in moist condition	13	5·5	0·9
Plants grown in subsoil stored in dry condition	12	5·6	0·8

The crops on pots 10 and 11 ought of course to weigh the same, and so should the crops on pots 8 and 9. The differences arise from several causes. No two plants are quite alike, nor any two lots of soil. Also, however carefully an experiment is made, and however skilful the operator, there are always errors in the work. These and other unavoidable troubles are lumped together in what is called the EXPERIMENTAL ERROR.

There is clearly an increase in crop as a result of storing the surface soil in a moist condition, showing that additional plant food has been *made* since these pots were put up. On the other hand, it does not appear that much plant food has been made in the subsoil during this time. Further evidence on this point is given by an experiment similar to that in Fig. 22, but where mustard is grown in *subsoil* kept moist, and uncropped for some time, and in *subsoil* previously cropped with rye. The results in grams were:

	Green weight	After drying
Mustard growing in subsoil previously cropped with rye	12·6	2·27
Mustard growing in subsoil previously uncropped	12·9	2·26

These should be compared with the figures on p. 36. Although the subsoil lay fallow for a long time it produced little if any plant food but seemed just as poor as the subsoil that has been previously cropped. These observations give us a clue that must be followed up in answering the next question.

WHAT HAS THE PLANT FOOD BEEN MADE FROM? Clearly it is not made from the sand, the clay or the chalk since all these occur in the subsoil. We have seen (chapter I) that the top soil differs from the subsoil in containing a quantity of material that will burn away and is in part at any rate made up of plant remains. Let us try to find out whether these remains furnish any appreciable quantity of plant food.

Fill one pot with surface soil, and another with the same weight of surface soil well mixed up with 60 grams of plant remains—pieces of grass, or stems and leaves of other plants cut up into small fragments. At the same time put up two pots of subsoil, one of which, as before, is mixed with 60 grams of plant remains, and also put up two pots of sand, one containing 60 grams of plant remains and the other none. Sow all six pots with mustard and keep watered and well tended. The result of one experiment is shown in Fig. 23 and the weights of the crop in grams were:

	Green weight	After drying
Top soil and pieces of plants (pot 6)	42·0	5·0
Top soil alone (pot 3)	17·7	2·6
Difference in top soil	24·3	2·4

	Green weight	After drying
Subsoil and pieces of plants (pot 7)	10·5	1·9
Subsoil alone (pot 4)	5·1	1·1
Difference in subsoil	5·4	0·8

Now let us look at these results carefully. The experiment with surface soil shows that the pieces of stem and leaf have furnished a good deal of food to the mustard and have caused a gain of 24·3 grams in the crop. If we knew what the pieces were made of we could push the experiment still further and find out more about plant food, but this involves chemical problems and must be left alone for the present. We can, however, say that plant remains are an important source of plant food, and since we suppose the black material of the soil to be made of plant remains (see p. 29), it will be quite fair to say also that this black material, the humus, is a source of plant food.

We can now explain some at any rate of the differences between the surface soil and the subsoil. The surface soil contains a great deal of the black material, which forms plant food, while the subsoil does not. Thus plants grow well on the surface soil and starve on the subsoil. We can also explain why gardeners and farmers speak of black soils as rich soils; they contain more than other soils of this black material that makes plant food. Still further, we can explain why the farmer often sows plants like mustard, tares or clover, and then ploughs them into the ground. They are not wasted, but they make food for the next crop that goes in.

Now turn to the results of the subsoil experiments. The leaves and stems have increased the crop, but only by 5·4 grams: they have not been nearly so effective as in the surface soil. It is evident that

Fig. 23. Pieces of grass, leaves, etc. change to plant food in the surface soil but not so well in the subsoil.

Pot 3: top soil alone.
Pot 6: top soil and pieces of plants.
Pot 4: subsoil alone.
Pot 7: subsoil and pieces of plants.

the mustard did not feed directly on the leaves and stems put in; if it had there should have been an equal gain in both cases. The leaves and stems clearly have to undergo some change before they are made into plant food and the soil has something to do with this change. After the crops are cut the soils should be tipped out and examined.

More of the original pieces of leaf and stem are found in the subsoil than in the surface soil. That is to say, there has been more change in pot 6 containing surface soil than in pot 7 containing subsoil. The 'something', whatever it may be, that changes plant remains like leaves, stems, pieces of grass, roots, etc. into plant food therefore acts better in the surface soil than in the subsoil. Here then we have another difference between surface and subsoils.

SUMMARY. The experimental results obtained in this chapter may now be summed up as follows:

(1) Plant food is present mainly in the top soil and to a much smaller extent in the subsoil.

(2) There is not much present even in the top soil, so little indeed that we could not detect it by weighing.

(3) It is, however, always being made in the top soil, if water is present. Only little is made from the subsoil.

(4) The remains of leaves, stems, roots, etc., furnish an important source of plant food.

(5) But they have first to undergo some change, and the agent producing this change is more active in the top soil than in the subsoil.

(6) The top soil is much the most useful part of the soil and should never be buried during digging or trenching, but always carefully kept on top.

CHAPTER VII

THE DWELLERS IN THE SOIL

APPARATUS REQUIRED. Garden soil. Bottle and cork. Two Erlenmeyer flasks, 50 c.c. capacity. Cotton-wool. Milk (about half a pint). Leaf gelatine. Soil baked in an oven. Three saucers. The apparatus in Fig. 28 (two lots). Wash bottle containing lime water (Fig. 27, also p. 15).

In digging a garden a number of little animals are found, such as earthworms, beetles, ants, centipedes, millipedes and others. There are also some very curious forms of vegetable life. By carefully looking about it is not difficult to find patches of soil covered with a greenish slimy growth; they are found best under bushes where the soil is not disturbed, or else where the soil has been pressed down by a footmark and not touched since. Any good soil left undisturbed for a time shows this growth.

Put some fresh moist garden soil into a bottle and cork it up tightly so that it keeps moist. Write the date on the bottle and then leave it in the light where you can easily see it. After a time—sometimes a long, sometimes a shorter time—the soil becomes covered with a slimy growth, greenish in colour, mingled here and there with reddish brown. The longer the soil is left the better. Often after several months something further may happen; little ferns begin to grow and they live a very long time. At Rothamsted a bottle of soil was put up just like this in 1874. Before long a beautiful fern started growing inside the bottle, and it remained healthy and vigorous for over forty years till the bottle was accidentally broken in 1916. If, instead of being kept moist, the rich garden soil is left in a dry shed during the whole of the winter so that it gradually loses its moisture, it generally shows quite a lot of white fluffy growth.

All of these living things are very wonderful, and some, especially earthworms, are very useful to gardeners and farmers.

After a shower of rain, look carefully in the garden or else on a lawn, common, or pasture field where the grass is closely grazed by farm animals or does not naturally grow long, and you will find numbers of tiny heaps of soil scattered about. Carefully brush away a heap and a little hole is seen, now hit the ground near it a few times with a stick or stamp on it with your foot, and the worm, if he is near the top, comes up. When he is safely out of the way dig carefully down with a knife or trowel so as to examine the hole or 'burrow'. At the top you generally find it lined with pieces of grass or leaves that the worm has pulled in; lower down the lining comes to an end, but the colour of the burrow is redder than that of the rest of the soil wherever the soil has a greenish tinge. These holes are useful because they let air and water down into the soil.

The following experiment shows what earthworms can do. Fill a pot with soil from which all the worms have been carefully picked out and another with soil to which earthworms have been added, one worm to every pound of soil. Leave the pots out of doors where the rain can fall on to them. You can soon see the burrows and the heaps of soil or 'casts' thrown up by the worms: these casts wash or blow over the surface of the soil, continually covering it with a thin layer of material brought up from below. Consequently the soil containing earthworms always has a fresh clean look. After some time the other soil becomes very compact and is covered with a greenish slimy growth. When this happens carefully turn the pots upside down, knock them so as to detach the soil and lift them off. The soil where the earthworms had lived is full of burrows and looks almost like a sponge. Fig. 24 shows what happened in an experiment lasting from June to October. The other soil where there were no earthworms shows no such burrows and is rather more compact than when it was put in.

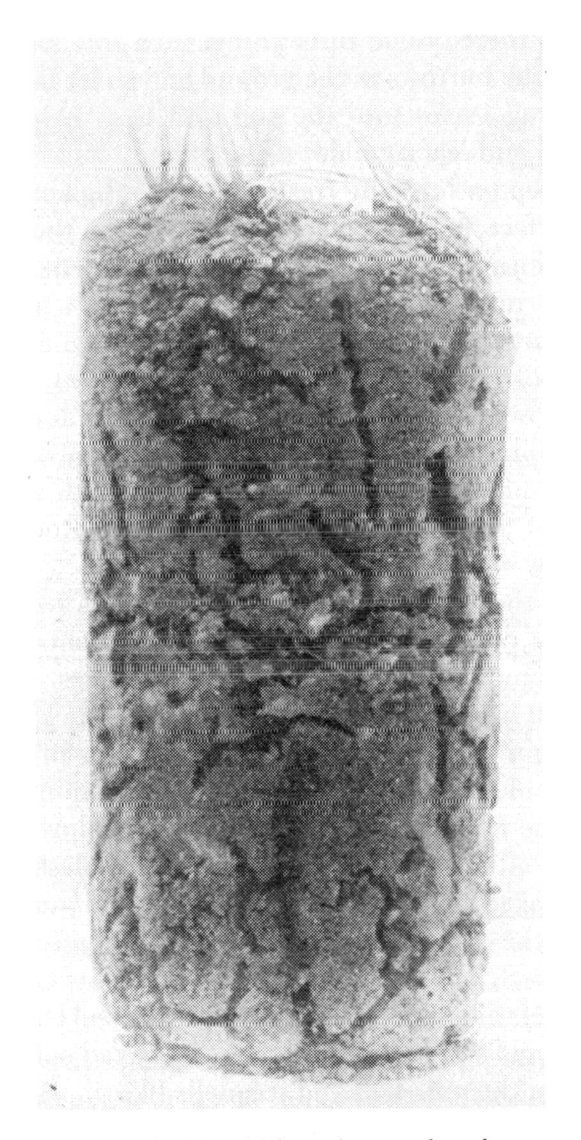

Fig. 24. Soil in which earthworms have been living and making burrows.

Earthworms therefore do three things:

(1) They make burrows in the ground and so let in air and water.

(2) They drag leaves into the soil and thus help to make the mixture of soil and leaf mould.

(3) They keep on bringing fresh soil up to the surface, and they disturb the surface, keeping it clean and free from the slimy growth.

All these things are very useful, and so a gardener should never want to kill worms. The great naturalist, Darwin, spent a long time in studying earthworms at his home in Kent and wrote a very interesting book about them, called *Earthworms and Vegetable Mould.* He shows that each year worms bring up about one-fiftieth of an inch of soil, so that if you laid a penny on the soil now and no one took it, in fifty years it might be covered with an inch of soil. Pavements that were on the surface when the Romans occupied Britain are now covered with a thick layer of soil.

But besides these there are some living things too small to see, that have only been found by careful experiments, but you can easily repeat some of these experiments yourselves. Divide a little rich garden soil into two parts and bake one in the kitchen oven on a patty tin. Pour a little milk into each of two small flasks, stop up with cotton-wool (see Fig. 25) and boil for a few minutes very carefully so that the milk does not boil over, then allow to cool. Next carefully take out the stopper from one of the flasks and drop in a little of the baked soil, label the flask 'baked soil' and put back the stopper. Into the other flask drop a little of the untouched soil and label it; leave both flasks in a warm place till the next day. Carefully open the stoppers and smell the milk: the baked soil has done nothing and the milk smells perfectly sweet; the unbaked soil, on the other hand, has made the milk bad and it smells like poor cheese. If you have a good microscope you can go further: look at a drop of the liquid from each flask and you find in each case the round fat globules of the milk, but the bad milk contains in addition some tiny creatures, looking like very short pins, darting in and out among the fat globules. These living things must have come from the unbaked soil or they would have been present in both flasks: they must also have been killed by baking in the oven.

Another experiment is easy but takes a little longer to show. Mix two sheets of leaf gelatine with a quarter of a pint of boiling water, pour into each of three saucers, and cover over with plates. Then stir up some baked soil in a cup half full of cold boiled water, and

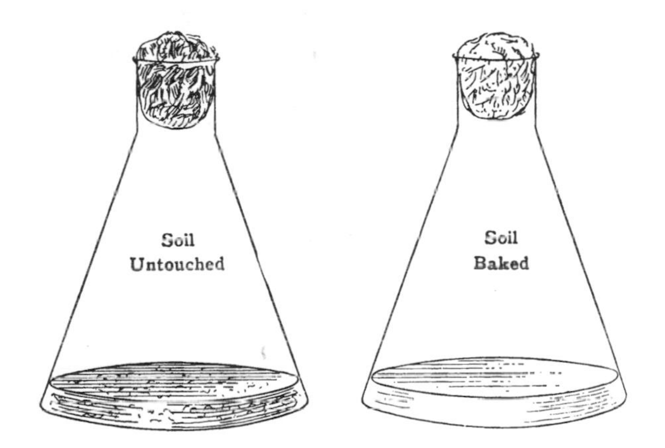

Fig. 25. Fresh soil turns milk bad, but baked soil does not.

quickly put a teaspoonful of the liquid into a second cup, also half full of cold boiled water. Stir quickly and put a spoonful on to the jelly, tilting it about so that it covers the whole surface and label the saucer 'baked soil'. Do the same with the 'unbaked soil', labelling the saucer; leave the third jelly alone and label it 'untouched'. Cover all three with plates and leave in a warm place. After a day or so little specks begin to appear on the jelly containing the unbaked soil, but not on the others (Fig. 26); they grow larger, and before long they change the jelly to a liquid. The other jellies show very few specks and are little altered. These creatures, making the specks, came from the soil because so few are found on the jelly alone; they were killed in the baking and so do not occur on the baked soil jelly.

You can also show that breathing is going on in the soil even after you have picked out every living thing that you can see. First of all you must do a little experiment with your own breathing so

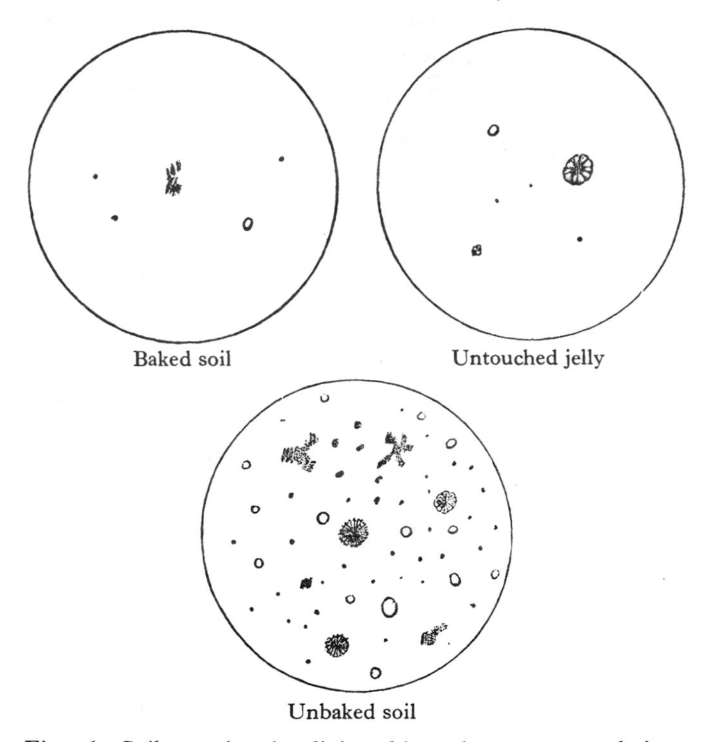

Fig. 26. Soil contains tiny living things that grow on gelatine.

that you may know how to start. Shake up a little fresh lime with water and leave it to stand for 24 hours. Pour a little of the clear liquid into a flask or bottle fitted with a cork and two tubes, one long and one short like that shown in Fig. 27. Then breathe in through the tube *A* so that the air you take in comes through the lime water: notice that no change occurs. Next breathe out through the tube *B* so that your breath passes through the lime water; this time the lime water turns very milky. You therefore alter in some way the air that you breathe: you know also that you need fresh air.

Now we can get on with our soil experiments. Take two small flasks of equal size fitted with corks and joined by a glass tube bent like a U with the ends curled over. Put some lime water into each flask and a little water in the U-tube. Now make a small muslin bag

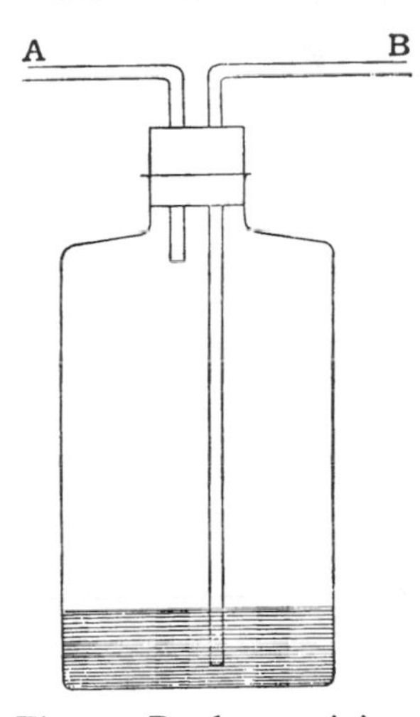

Fig. 27. Bottle containing lime water, used to show that breath makes lime water milky.

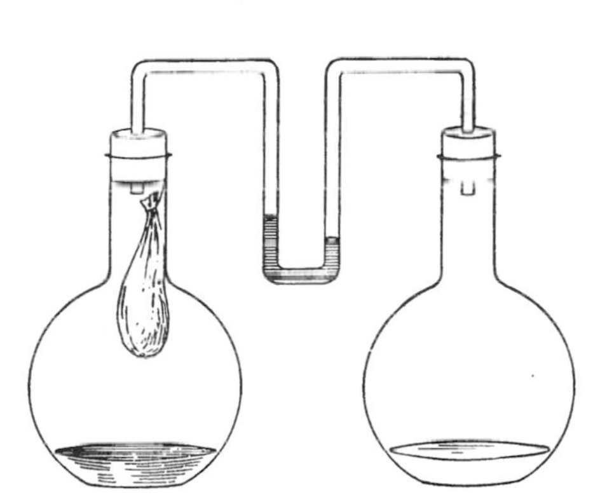

Fig. 28. A bag of soil is fixed into a flask containing lime water. In a few days some of the air has been used up, and the lime water has turned milky.

like a sausage: fill it with moist fresh garden soil, tie it up with a silk thread and hang it in one of the flasks by holding the end of the thread outside and pushing in the cork till it is held firmly (see Fig. 28). Fix on the other flask, and after about 5 minutes mark the level of the liquid with a piece of stamp paper; leave in a warm place but out of the sun.

In one or two days you will see that the water in the U-tube has moved towards the soil flask, showing that some air has been used up by the soil; further, the lime water has turned milky. But in the other flask, where there is no soil, the lime water remains quite clear.

This proves, then, that some of the tiny creatures want air just as much as we do. The air reaches them through passages in the soil, through the burrows of earthworms and other animals, or by man's efforts in digging and ploughing.

Now try the experiment with very dry garden soil: little or no change takes place. As soon as you add water, however, breathing begins again, air is absorbed and the lime water turns milky just as before. Water is therefore wanted just as much as air.

Food also is needed. You can show this by making up two heaps of waste vegetable material: leaves, cabbage stalks, long grass or weeds, lawn mowings, etc. Make the heaps as nearly equal as you can, then take one to pieces and sprinkle it with the proper quantity of compost starter: Adco, or other good preparation that can be had from a horticultural dealer, following the directions given. Remake the heap. Something soon happens. The heap gets hot and you may have to add water to prevent it drying out. The vegetable matter darkens and in a few weeks much of it has changed into mould; when the heap is turned and made up again it is well on the way to become what gardeners call COMPOST. But the heap without the starter changes much less; it does not become so hot or produce so much mould.

The starter is really food for the tiny living things that break down the leaves and stems and make humus: it is the same as nitrogenous and phosphatic plant food (p. 84). Other forms of the same foods also encourage these tiny workers: instead of the starter you could use farmyard manure, liquid manure, etc.

But you do not need to add the tiny workers, there are always plenty of them about, and when they have food and water and the other things they want many of them multiply by simply splitting into two.

If you had very magnifying eyes and could see things so enlarged that these little creatures seemed to you to be an inch long, and if you looked down into the soil, it would seem to you to be an extraordinarily wonderful place. The little grains of soil would look like great rocks and on them you would see creatures of all shapes and sizes moving about, and feeding on whatever was suitable to them, some being devoured by others very much larger than themselves, some apparently dead or asleep, yet waking up whenever it becomes warmer or there was a little more moisture. You would see them changing useless dead roots and leaves into very valuable plant food; indeed it is they that bring about the changes observed in the experiments of chapter VI. Occasionally you would see a very strange sight indeed—a great snake-like creature, over 3 miles long and nearly half a mile round, would rush along devouring everything before it and leave behind it a great tunnel down which a mighty river would suddenly pour, and what do you think it would be? What you now call an earthworm and think is 4 inches long, going through the soil leaving its burrow along which a drop of water trickles! That shows you how tiny these little creatures are.

These busy little creatures are called MICRO-ORGANISMS because of their small size. But they are not all useful. Some can turn milk bad as we have already seen, and therefore all jugs and dishes must be kept clean lest any of them should be present. Others can cause disease to plants, human beings and animals. Finger and toe in cabbages and turnips, root rots, and many other plant diseases are brought about by certain soil organisms. It has happened that a child who has cut its finger and has got some soil into the cut, and not washed it out at once, has been made very ill. You may sometimes notice sheep limping about in the fields, especially in damp fields; an organism gets into the foot and causes trouble.

We have seen that earthworms, of which there are several different kinds, are beneficial, but there are other soil animals that

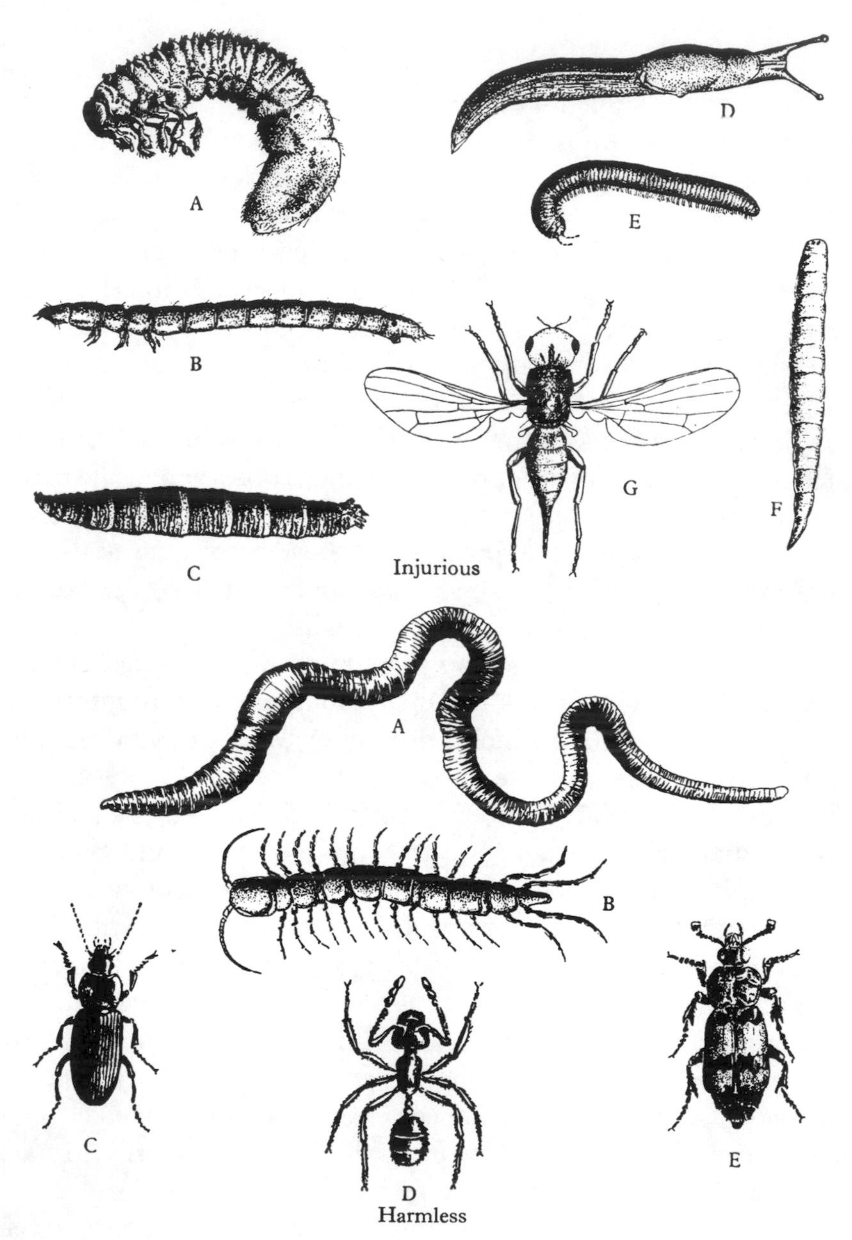

Fig. 29. Above: *A*, chafer grub; *B*, wireworm; *C*, leatherjacket; *D*, slug; *E*, millipede; *F*, carrot-fly larva; *G*, carrot-fly. (*A*, *B*, *C*, *D*, *E*, natural size, *F* and *G*, 5 times natural size.) Below: *A*, earthworm; *B*, centipede; *C*, carabid beetle; *D*, ant; *E*, burying beetle. (*A*, *B*, *C* and *E*, natural size; *D*, 5 times natural size.)

are not. Some feed on plant roots or stems and may be very destructive. Wireworms, leatherjackets, millipedes (one kind is pale yellow with a double row of crimson or purple spots, another kind is black, but both are round), slugs and snails are big enough to be seen: they can be trapped and destroyed. But eelworms, just too small to be seen, may do even more harm to potatoes and sugar beet. On the other hand centipedes are useful because they feed on other soil animals: they are flat and move quickly and should not be destroyed. You should collect and label the various soil insects you find. Fig. 29 shows some common ones; the seven in the upper part are injurious, the five in the lower part are beneficial or harmless.

SUMMARY. The soil is full of living things, some large like earthworms, others very small. Earthworms are very useful: they make burrows in the soil, thus allowing air and water to get in: they drag in leaves and they keep on covering the surface with soil from below. Besides these and the other large creatures, there are micro-organisms so small that they cannot be seen without a very good microscope: they live and breathe and require air, water and food. Some are very useful and change dead parts of plants or animals into valuable plant food. Almost anything that can be consumed by fire can be consumed by them. Others are harmful.

CHAPTER VIII

THE SOIL AND THE PLANT

APPARATUS REQUIRED. Dry powdered soil, sand, clay, leaf mould, seeds. Three funnels, disks, stands and glass jars. Two glass tubes about half an inch diameter and 18 inches long. Muslin, string, three beakers. Three lamp chimneys standing in tin lids. Pot experiments (p. xvi), growing plant. Two test-tubes fitted with split corks (Fig. 36). Soil acidity indicator.*

If you have ever tried to grow a plant in a pot you must have discovered that it needs much attention if it is to be kept alive. It wants water or it withers; it must be kept warm enough or it is killed by cold; it has to be fed or it gets yellow and starved; also it needs fresh air and light. These five things are necessary for the plant:

Water	Fresh air
Warmth	Light
Food	

We may add two others: enough depth of soil, and absence of harmful substances.

Wild plants growing in their native haunts get no attention and yet their wants are supplied. We will try and find out how this is done.

Water comes from the rain, but the rain does not fall every day. How do the plants manage to get water on dry days? A simple experiment will show you one way. Put about four tablespoonfuls of dry soil on to the funnel shown in Fig. 30 and then pour on two tablespoonfuls of water. Measure what runs through. You will find it very little; most of the water sticks to the soil. Even after several days the soil was still rather moist. Soil has the power of keeping a certain amount of water in reserve for the plant, it only allows

* A coloured liquid obtainable at large chemists. There are several makes: B.D.H. and others, and full instructions are given.

a small part of the rain to run through. Do the experiment also with sand, powdered clay, and leaf mould. Some water always remains behind, but less in the case of sand than in the others. In one experiment 30 cubic centimetres of water were poured on to 50 grams of soil but only 10 cubic centimetres passed through, but when an equal amount was poured on to 50 grams of sand no less than 20 cubic centimetres passed through. Very sandy soils, therefore, possess less power of storing water than do soils with more clay or mould in them, such as loams, clays or black soils.

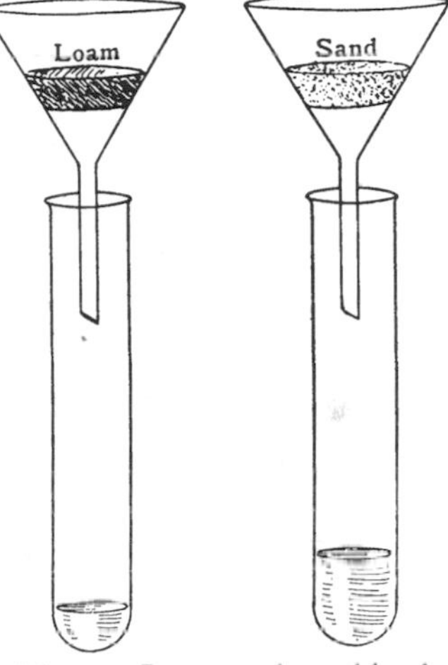

Fig. 30. Loam and sand both retain water, but loam better than sand.

Further, water has a wonderful power of passing from wet places to drier places in the soil. When you are watering on a dry day you will see that the water does not stay just where it falls but seeps along the soil and through it. You cannot make water run uphill but the soil can. Tie a piece of muslin over the end of a tube and fill with dry soil, tapping it down as much as you can, then stand the tube in water as in Fig. 31. Fill another with coarse sand and place in water. Notice that the water at once begins to rise in both tubes; it usually goes higher in the soil than it does in the sand. Enough water may pass up the tube in this way to supply the needs of a growing plant. Fill a glass lamp chimney with dry soil, packing it down tightly, put into water and then sow with wheat. The plants grow very well. A longer tube may be made from two chimneys fastened together by means of a tin collar stuck on with Canada balsam or sealing wax. In our experiment (Fig. 32) the plants grew well in this also, but on a sandier soil, where the water could not rise so high, it might happen that they would not.

When plants are grown in pots, or seedlings are being raised in boxes, it is better to water them from below in this way than to do it from above. This kind of movement does not take the water far, however, and it is important only when the water level is near the surface.

There are great differences in the moisture content of various soils. In some districts there is much more rain than in others, and therefore the soils get a larger supply of water. Sandy soils allow water to run through while a loam holds it like a sponge. Further, water runs down hills and collects in low-lying hollows or valleys; here, therefore, the soil is moister than it is somewhat higher up. What will be the effect of these moisture differences on plants?

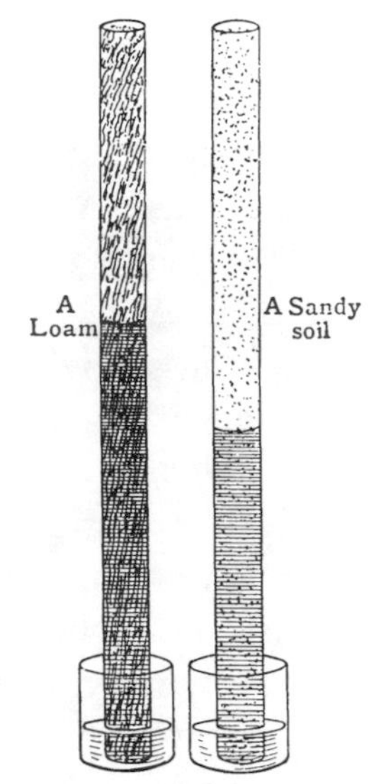

Fig. 31. Water can pass from wet to dry places in the soil; it can even travel upwards.

You must find out in two ways. Visit a soil that you know is dry—a sandy, gravelly or chalky soil in a high situation—and look carefully at the plants there, then go to some moister, lower ground and see what the plants show. You cannot be quite certain, however, that anything you see is simply due to water supply, because there may be other differences in the soil as well. So you must try the second method, and that is to find out by experiments what is the effect of varying quantities of water on the plant growth. Both methods must be used, but it may be more convenient to start the experiments first, and while they are going on to collect observations in your rambles.

Fill four glazed pots with dry soil: keep one dry; one only just moist; the third is to be very moist and should be watered more frequently than the second; and the fourth is to be kept flooded with water, any way out being stopped up. Sow wheat or mustard in all four and keep them out of the rain. The result of one experiment with mustard is shown in Fig. 33. Where no water was supplied there was no growth and the seeds remained unaltered. Where only little

Fig. 32. Plants growing in soils supplied from below with water.

water was supplied (pot 16) the plants made some growth, but not very much: the leaves were small and showed no great vigour; where sufficient water was given (pot 3) the plants grew very well and had thick stems and large leaves; where too much water was given (pot 15) the plants were very sickly and small.

The weights were:

	Green weight	After drying
Plants with too much water	3·9	0·5
Plants with too little water	5·3	0·9
Plants with a nice quantity of water	17·7	2·6

Fig. 33. Mustard growing in soils supplied with varying quantities of water.

Fig. 34 shows two pots of wheat, one kept only just sufficiently moist for growth, the other kept very moist but not too wet. You see what a difference there is; in the drier pot (*B*) the leaves are rather narrow and the plants are small, in the moister pot (*A*) the leaves are wide and the plants big. But there was also another difference that the photograph does not bring out very well—the plants in the rather dry soil were, as you can see, in full ear, ripe and yellow, while those in the very moist soil were still green and growing.

We see then (1) that on moist soils there is greater growth than on dry soils, but the plants do not ripen so quickly; (2) in very wet soils mustard—and many other plants also—will not grow.

This explains why wheat and barley, farm and garden seeds, all of which need to be quite ripe before they are harvested, are much

(*A*) Moist soil (*B*) Dry soil

Fig. 34. Wheat growing in moist and in dry soils.

more widely grown in the drier eastern half of England than in the wetter west, and also why the growers avoid clay soils.*

Water is not itself harmful. It is easy to grow many plants in water containing the proper food, but *air must be blown through the water at frequent intervals*. In the waterlogged soil of pot 15 the trouble arose not from too much water but from too little air. Air is wanted because plants are living and breathing in every part, in the roots as well as in the leaves.

* Except for wheat.

Now turn to what you have seen in your walks. You would probably notice that on the drier, sandy or gravel ground there was nothing like as great a growth of grass or of other plants as on

Fig. 35*a*. Plants found on a dry soil had narrow leaves, those on a moist soil had wider leaves.

Left–right, Broom (*Cytisus scoparius*); sheep's fescue (*Festuca ovina*); gorse (*Ulex europaeus*).

the moister soil. This is so much like what we found in the pot experiments that we shall not be wrong in supposing that the difference in water supply largely accounted for the difference in growth. But you may also have noticed something else. Plants in drier soil have generally narrow leaves and the grasses are rolled up and fine, whilst those on the damp soil, including the grasses,

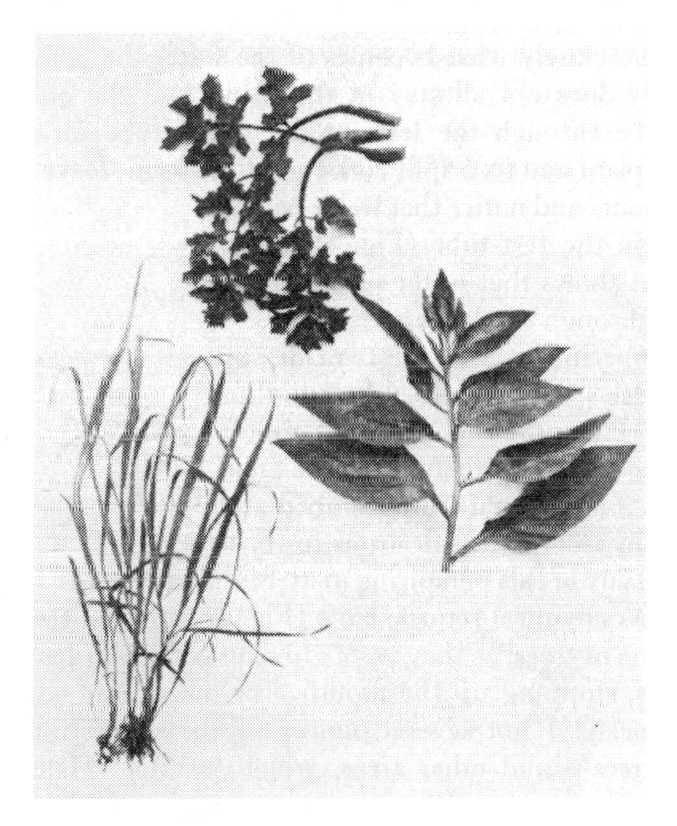

Fig. 35*b*. Plants found on a dry soil had narrow leaves, those on a moist soil had wider leaves.

Left, Wood brome-grass (*Bromus asper*); *top*, hogweed (*Heracleum sphondylium*); *right*, comfrey (*Symphytum officinale*).

have usually broad leaves. Thus in the dry sandy soil you may find broom, spurrey, sheep's fescue, pine trees, all with narrow leaves; whilst on the moister soil you may find burdock, primroses, cocksfoot and other broad-leaved plants. Figs. 35*a* and *b* show some plants we found on a dry, gravelly patch on Harpenden common, and on a moist loam in the river valley below.

Before we can account for this observation, we must ascertain a little more closely what becomes of the water the plant takes up. It certainly does not all stay in the plant, and the only way out seems to be through the leaves. Put a test-tube on the leaf of a growing plant and fix a split cork round the stem: leave in sunlight for a few hours and notice that water begins to collect in the test-tube (Fig. 36). The experiment shows that water passes out of the plant through the leaves.

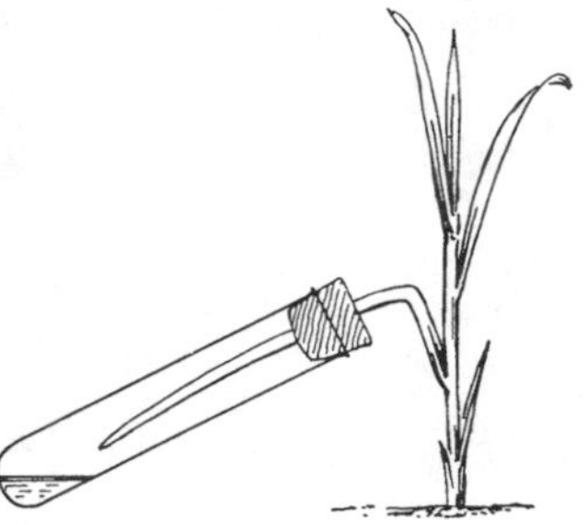

Fig. 36. Plants give out water through their leaves.

This experiment was first made by Stephen Hales, and described by him thus in 1727: 'Having by many evident proofs in the foregoing experiments seen the great quantities of liquor that were imbibed and perspired by trees, I was desirous to try if I could get any of this perspiring matter; and in order to do it, I took several glass chymical retorts, *b a p* [Fig. 37] and put the boughs of several sorts of trees, as they were growing with their leaves on, into the retorts, stopping up the mouth *p* of the retorts with bladder. By this means I got several ounces of the perspiring matter of vines, figtrees'—and other trees, which 'matter' Hales found to be almost pure water. The test-tube experiment should now be made with a narrow-leaved grass like sheep's fescue and with a wide-leaved grass like cocksfoot. You will find that wide-leaved plants pass out more water than those with narrow leaves, and hence wide-leaved plants occur in damp situations or on damp soils like loams and clays, while narrow-leaved plants can grow on dry, sandy soils.

Another thing you will notice is that fields lying at the side of a river and liable to be flooded, and fields high up in wet hill districts, are covered with grass mingled with clover and other plants: it is called 'grass land' for short. In a clay country there is also a great deal of grass land and much less ploughed land; if you live where there is much clay you can easily discover the reason. Clay becomes very wet and sticky when rain falls, and very hard in dry weather: it is, therefore, difficult to cultivate. Farmers cannot afford to spend too much money on cultivation, and so they prefer

Fig. 37. Stephen Hales's experiment in 1727.

grass, because once it is established it can go on indefinitely and need not be ploughed up and re-sown. And besides, farmers have learned by experience that grass can tolerate more water and less warmth than most other English crops. There is much more grass land in those parts of England where the rainfall is high and the temperature rather low—e.g. the northern parts of England—than in the eastern counties where the rainfall is low.

The difference in water supply, therefore, leads us to expect the following differences between dry sandy soils and moist clays or loams:

On sandy soils (the water content being small) the wild plants and trees usually have small leaves. Cultivated plants do not generally give very heavy crops, but they ripen early.

On clay soils (the water content being good) wild plants and trees usually have larger leaves. Cultivated plants give good crops, but they ripen rather late. If the water content is too high or the clay is too sticky the land is generally put into grass.

However, the differences are not always as great as you might expect. Plants have a wonderful way of adapting themselves to their surroundings and making the best of their conditions. On a sandy soil they produce long fibrous roots (unless there is some hindrance, like a hard layer, or acids that lime would put out of action), and so are able to search thoroughly for whatever water there is. On a clay soil their roots are much shorter. You should test this by growing plants in pots of sandy soil and of clay soil, and, when they are half-grown tip out the soil into a bucket of water and after it is thoroughly soaked, gently lift out the plants. Further, plants are able to take more of the water contained in a sandy than in a clay soil. So although a sandy soil is usually drier than a clay, plants are able to get more of whatever water is present.

This good root action in sandy soil explains why gardeners use sand or light compost for striking cuttings.

Plants require to be sufficiently warm. Some like tropical heat and can be grown only in hot-houses; others can withstand a certain amount of cold and will grow up on the mountains. Our common cultivated crops come in between and will not grow in too cold or exposed a situation; thus you find very little cultivated land 800 feet above sea-level, and not usually much above 500 feet. At this height it is left as grass land, and higher up as woodland, moor, or waste land. Grass requires less warmth and can therefore grow at greater heights than many other crops. If you start at the top of a hill in Wales, or Derbyshire, or any other high-lying region, and walk down, you will see that the top is moorland, lower down comes grassland, or wood land, if the slope is steep, still lower you may find arable land, and if the valley is damp you will find more grass at the

Fig. 38. Hill slope near Harpenden showing woodland at the top and arable land lower down.

bottom. Figs. 38 and 39 show typical views of the hill slopes farther south: they are taken near Harpenden. The top of the hill in each case is over 400 feet above sea-level, and has never been thought worth cultivating, but has always been left as wood because it is too exposed for farm crops and the soil is poor. On the lower slopes the arable fields are seen, while at the bottom bordering the river is rough grass land shown in Fig. 39. The top is too cold and windy, and the bottom too wet to be worth cultivating. Fig. 40 shows another common sequence: it is a picture of the Rheidol valley above Aberystwyth. The hill tops are bare, the slopes wooded, and the flat valley land is drained so that it is good for farming.

In winter and spring time the warmest part of Great Britain is Cornwall, especially the Penzance region. Here therefore are grown

Fig. 39. View farther along the valley, woodland and arable above, rough grass land near the river.

the earliest vegetables and flowers. Fig. 41 shows a field of daffodils in full bloom in January 1949; the flowers are being gathered for the London market.

As the plant root is alive it wants air. The effect of keeping air out can be seen by sowing some barley or onion seeds in the ground and then pouring a lot of water on and plastering the soil down with a spade. Sow another row in nicely crumbled soil, not too wet, press the seeds well in, but do not plaster the soil. This second lot will generally do much better than the first. If the ground round a plant is frequently trodden so that it becomes very hard the plant makes much less growth than if the soil were kept nice and loose. A good

Fig. 40. The Rheidol valley between Aberystwyth and the Devil's Bridge: the high ground bare, the slopes wooded, and the flat valley land farmed.

gardener takes very great pains in preparing his ground before he sows his seeds, and he is careful that no one should walk on his beds lest his plants should suffer.

HARMFUL SUBSTANCES IN THE SOIL. In dry countries salts are liable to occur in the soil and they do much damage to vegetation: you have heard of the salt round the Dead Sea, and of the Salt Lake in Utah. In this country we are saved from this trouble by our rainfall, but in parts of Wales lead or zinc ores from mines have got on to the soil and poisoned the plants. The commonest harmful substances in British soils are acids formed in very wet regions where rain has washed the lime out of the soil, or where the original rock material (p. 102) contained little lime: e.g. some of the dry sands; on such soils

Fig. 41. Picking daffodils in January in the warm Scilly Isles.

the brassicas (cabbages, sprouts, turnips, etc.) in the gardens get 'finger and toe', and on farms mangolds, sugar beets and clover fail to grow up. The remedy is to add lime.

Test the soil in the school garden with the Soil Acidity Indicator, put a few drops of the liquid on to the soil as directed, and see by the change in colour whether the soil is acid or not. If it is acid measure out a square, 2 yards each side, divide it into four equal plots, to one add 4 oz. of lime very evenly—best by stirring it in water and watering it in; to the plot diagonally opposite add equally carefully 1 lb. of lime. The other two plots are to remain without lime. Leave for some time so that the rain may wash the lime well into the soil, then test again with the indicator. This experiment is best done in early spring. Then plant brassicas or sow barley evenly over all four plots and watch the growth; afterwards examine the roots.

If your soil is not acid the lime may have no very marked effect. The county agricultural officers will test farm soils and advise whether lime is needed or not.

Some plants will put up with acidity better than others. Potatoes do not like lime, nor do some of the garden plants such as many lilies, heaths, azalias, rhododendrons; indeed gardeners will tell you that these hate lime. Some wild plants also do not like it: Mayweed, spurry, on light soils; sorrel, foxgloves on heavy soils: if you see these growing well the soil is likely to be acid.

SUMMARY. We may now collect together the various things we have learnt in this chapter. Plants require water, air, warmth, food and light, and they will not grow if harmful substances are present. The rain water that falls remains for some time in the soil, and does not at once run away or dry off: water can also move from wet to dry places in the soil. Also the plant roots can grow in all directions; they do not stay in one place waiting for water to come to them. Therefore the plant does not need rain every day, but can draw on the stock in the soil during dry weather. A sandy soil is usually drier than a loam or a clay, especially if it lies rather high: plants growing on a sandy soil make less growth and have narrower and smaller leaves than those on a moister soil, but they ripen earlier.

Situations more than five or six hundred feet above sea-level are, in England, as a rule, too bleak and exposed for the ordinary cultivated crops. Such land is, therefore, either grass land, moorland, downland or woodland.

The roots of plants are living and require air. The soil must not be trodden too hard round them or air cannot get in, nor can it if too much water is present.

Grass can put up with more water and less warmth than most cultivated crops.

Instances of these facts may be found in going down any hill 500 feet or more in height: the top is usually wood, poor grazing, or waste, being too cold for crops, below this may come grass land, lower still arable land. It is both warmer and moister in the valley (since water runs down hill), and so we can account for the proverbial fertility of valleys. But just near the river, if there is one, the ground may be too wet for crops, and therefore grass is grown. Clay land that is rather too wet to plough is usually left in grass.

CHAPTER IX

SOIL MANAGEMENT: CULTIVATION, MANURING AND LIMING

APPARATUS REQUIRED. Plot experiments, hoeing and mulching. Factory thermometer with stem 6 inches long. Soil sampler (Fig. 44, p. 78). This tool consists of a steel tube 2 inches in diameter and 9 inches long, with a slit cut along its length and all the edges sharpened. The tube is fixed on to a vertical steel rod, bent at the end to a ring 2 inches in diameter, through which a stout wooden handle passes. It is readily made by a blacksmith.

The previous chapter shows what the plant needs from the soil. We shall now try to find out how these needs can be supplied so that plants may make better growth.

Farmers and gardeners throughout the spring, summer and autumn, are busy ploughing or digging, hoeing or in other ways cultivating the soil. Unless all this is well done the soil fails to produce much; the sluggard's garden has always been a by-word and a reproach. In trying to understand why they do it we must remember that the seed must have fine soil pressed round it, so that the young roots can strike straightway into the soil and not just into the air; also that plant roots need water, warmth and air; if the soil is too compact or if there is too much water the plant suffers, as we have seen.

One great object of cultivation is, therefore, to make the soil fine and then to prevent it becoming too compact and too wet. After the harvest the farmer breaks up his ground with a plough and then leaves it alone till seed time. A gardener does the same thing with a spade and he should dig manure into the lower spit so as to deepen the soil. If there is frost during the winter both farmer and gardener are pleased because they say the frost 'mellows' the ground; you can see what they mean if you walk on a frosty morning over a ploughed field. The large clods of earth are no longer sticky, they already show signs of breaking up, and if they are not frozen too hard can easily be shattered by a kick. The change has been brought about in exactly the same way as the bursting of water-pipes by frost. When water freezes it expands with enormous force and bursts open anything that confines it; water freezing in the pores of the soil forces the little fragments apart. This action is so important that further illustrations should be looked for. A piece of wet chalk left out on a frosty night often crumbles to pieces. It is dangerous to climb cliffs in the early spring because pieces of rock that have been split off during the winter frosts by the expanding water may easily give way. Frost plays havoc with walls built of flints and with old bricks that are beginning to wear. If there are several frosts, with falls of rain or snow and thaws coming in between, the soil is moved about a good deal by the freezing and melting water. Bulbs and cuttings are sometimes forced out of the ground, whilst grass and young wheat may be so loosened that they have to be rolled in again as soon as the weather permits. When the ground has been dug in autumn and left in a very rough state all this loosening work of the frost is very much helped, because so much of the soil may become frozen. If in spring you dig a piece of land that has already been dug in autumn, and then try digging a piece that has not, you will find the first much easier work than the second in all but very sandy soils.

A short time before the seeds are sown, the soil has to be dug or cultivated again so that it may become more level and broken into smaller pieces. The farmer then harrows and the gardener rakes it, and it becomes still finer. Very great care is bestowed on the preparation of the seed-bed, and it will take you longer to learn this than any other part of outdoor gardening. The soil has to be made fine and dry, and no pains must be spared in getting it so.

When at last the soil is fine enough the seed is put in. But it is not enough simply to let the seed tumble into the ground. It has to be pressed in gently with a spade or a roller, not too hard or the soil becomes too sticky. Then the soil should be left alone.

Fig. 42. Making a firm seed-bed and sowing grass seed.

Farmers now use tractors and big implements which do this work both quickly and well, and on light soils can often combine several operations (Figs. 42 and 43).

If you watch an allotment holder who grows onions really well working away at his seed-bed you will see what a beautifully fine TILTH he gets. If you try to do the same you will probably fail; his seeds will be up before yours and will grow into bigger and healthier plants. Only after long practice will you succeed, and then you will have mastered one of the great mysteries of gardening.

As soon as the plants are up they have to be hoed, and while they are young this must be done whenever weeds appear or whenever

Fig. 43. Ploughing, sowing and harrowing in seeds on a Wiltshire farm.

the soil surface becomes hard and glazed. You can see the effects of hoeing by making an experiment in the school garden in late spring or early summer. Set out four plots, each 6 feet square, hoe off the weeds. Then leave one entirely alone; keep the second free from weeds by cutting them off with a spud directly they appear, but do not disturb the soil more than you can possibly help; hoe the third once a week; and the fourth three times a week so far as weather permits. Label the plots so that there can be no mistake.

Remembering what was said about the 'error of the experiment' on p. 39, set out alongside of these another group of plots, but arranged in a different order. Take notes of what happens in different weather conditions. On light soils there may be little difference between the plots, but on heavy soils the hoed soils will allow rain water, especially from a downpour, to soak in more easily than the unhoed soil, where some of the water may run off in a muddy stream carrying soil with it, and some may remain in little pools on the surface; both are bad for the plants, but the gentle soaking in is good for them. When the soil dries the surface of the unhoed plots looks glazed and becomes hard, but you will break up this crust on the hoed plots. In hot dry weather the unhoed plot, if the soil is heavy, begins to crack and the cracks spread downwards. On the hoed plots, however, there is no cracking. If weeds have started to grow on the untouched plot you may see how the roots get exposed or even torn by these cracks; you can also find examples in the fields. This cracking is of course bad for the plants, and hoeing helps them by preventing it.

These differences in surface conditions generally affect the temperature of the soil: they did in the St George's School garden, and you should see if they do in yours. Take the temperatures by placing the thermometer: (*a*) on the surface of the plot, (*b*) ½ to 1 inch down, (*c*) 3 inches down, (*d*) 6 inches down; but first make a hole with a pencil or round stick just big enough to hold the thermometer, otherwise you may break it. The soil must be pressed well against the thermometer. We used the long-stemmed form called a factory thermometer which is easier to read than the ordinary form. Enter the results as in the table, which shows what happened in the St George's School garden.

Date	Air temperature °C.	Depth (inches)	Soil temperature (degrees Centigrade)		
			Unhoed	Hoed once weekly	Hoed three times weekly
20 June	30	½ to 1	35	31·5	31·5
		3	30·5	29·8	28·8
		6	27	26·5	24
27 June	18	½ to 1	17·5	17	17
		3	16·7	16·3	16·2
		6	15·8	15·5	15·5

Remarks. 20 June: hot sunny day; there had been no rain since 11 June. Note that the surface of the soil is hotter than the air. 27 June: cold, cloudy day; several cold, wet days during the past week.

On the cold day there was very little difference between the plots, but on the hot day the hoed plots were cooler than the others. Now only the top inch is touched by the hoe, and so it appears that the layer thus loosened shields the rest of the soil from the sun's heat. If this is the case, we ought to find that any other loose material would act in just the same way. We must, therefore, set out a fourth plot alongside the others, cover it with straw or cut grass (a cover like this is called a MULCH), and take the temperature there. Some of the results were as follows:

Date	Air temperature °C.	Depth (inches)	Soil temperature (degrees Centigrade)	
			Hoed plot	Mulched plot
24 September	15	½	17·5	12·25
		3	12·5	11·75
		6	12·25	11·5
5 October	17	½	17	15·5
		3	16·7	15
		6	15·5	14·5

Remarks. 24 September: warm day after a rather cold spell. 5 October: warm day after a long spell of dry, warm weather.

On a dry day remove the mulch; you will generally find the soil is moister than on the unmulched plots.

A slate or flat stone acts like a mulch; if you leave one on the soil for a few days in hot weather and then lift it up on a hot day you will see that the soil underneath is quite moist; you may also find several slugs or other animals that have gone there for the sake of the coolness and the moisture. Plants and trees also keep off the sun's heat. Grass land in summer and autumn, and even in early winter, is cooler near the surface than bare land. At Harpenden we found:

Date	Depth (inches)	Soil temperature (degrees Centigrade)	
		Grass land	Bare land
24 September	½	13	17·5
	3	12·5	12·5
	6	12·5	12·25
5 October	½	15·5	17
	3	15	16·7
	6	14·5	15·5

The sun's heat travels so slowly into the soil in summer that months may pass before it gets far down, but then, as it takes so long to get in, it also takes a long time to get out, and it takes still longer to get either in or out if there is a mulch or if grass is growing.

During the early winter you may notice that the first fall of snow soon melts on the bare arable land but remains longer on the grass; towards the end of the winter, however, the reverse happens and the snow melts first on the grass. There is no difficulty in explaining this. The bare land is, as we have seen, warmer in autumn and early winter than grass land, and so it melts the snow more rapidly. But during winter the grass land loses its heat more slowly, and therefore it is warmer at the end of the winter than the bare land, hence the snow melts more quickly.

It might be expected that hoeing or mulching, by keeping the soil cooler in hot weather, would also keep it moister. You can find out whether that is so on your plots. Take a sample* of the soil, weigh it, then set it to dry in a warm place and weigh again: the difference is the loss of moisture. Find out what it would be in a hundred parts

* The amount of soil you take may be more than you can weigh: in that case tip it on to a dry table, spread it out, pick up a teaspoonful here and there to make a fair sample and stop when you have got enough. *But you must work quickly!*

of the fresh soil. The sample may be taken by pushing the tool, shown in Fig. 44, 6 inches down into the soil. A trowel will do, but it is not so good, as the sides of the hole must be kept vertical and all samples must be taken to the same depth on all the plots. Usually the differences in moisture content are not very great, and many experiments have shown that the benefit of hoeing is something more than its effect on temperature and on moisture content.

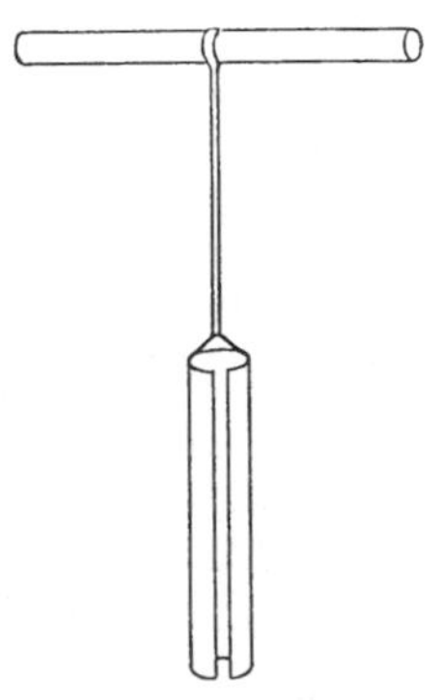

Fig. 44. Soil sampler. (See p. 71 for description.)

That 'something more' is the destruction of weeds, the breaking up of the glazed surface, and the prevention of cracks in the soil. Weeds are particularly harmful to plants, especially young plants, and should be cut out directly they appear, however little they may be. But in doing this be careful not to disturb the roots of your plants: when they begin to grow and to spread about in the soil you may easily damage them. Many experiments have shown that the best way of dealing with weeds is to scrape them off directly they appear without going deeply into the soil. Here is a good experiment that was made at Ottershaw College, Surrey. There were four sets of potato plots treated as follows:

(1) weeds kept down by scraping the top half inch of soil;

(2) two cultivations, 3 inches deep on one set, 6 inches deep on another; weeds appearing in the meanwhile were scraped off;

(3) two cultivations as above but no scraping: some weeds grew;

(4) five cultivations as above.

The potatoes were weighed, and the results calculated as tons per acre. They were:

(1) Weeds scraped, no cultivation	(2) Weeds scraped, two cultivations	(3) Weeds left, two cultivations	(4) Five cultivations
11·8	11·0	9·7	10·7

The experiment takes more time than you may be able to spare, but you can try a simplified form. At St George's School we sowed two rows of maize 2 yards apart, kept one hoed and left the other to

Fig. 45*a*. A hoed plot showing the maize without weeds.

struggle with the weeds as they grew up. Figs. 45 *a* and *b* show the result: the hoed plants grew nearly to the top of the pole, the others were only about half as high.

Before the war land sometimes went out of cultivation because of low prices: it was called DERELICT. It became covered with wild grasses and weeds, and gradually bushes grew up. In order to see what would happen, a part of the Broadback wheat field at Rothamsted was left unharvested in 1882 and the plot has not

Fig. 45*b*. An untouched plot showing the maize with many weeds.

been touched since; the wheat was allowed to shed its seed, and to grow up without any attention. Weeds flourished, but the wheat did not; the next year there was but little wheat, and by 1886 only a few plants could be seen, so stunted that one would hardly recognize them. The ground still remains untouched; it soon became a dense thicket and is now a wood (Fig. 46). Most of our land would become like this if it were neglected for a few years.

Farmers occasionally leave their ground without a crop for some months in spring and summer, and cultivate it when necessary to

Fig. 46. A plot of wheat left untouched since 1882 at Rothamsted has now become a wood.

kill the weeds. This practice is called 'fallowing', and is very ancient; it is much less common now that crops like mangolds and swedes are grown, which can be hoed during the summer if weeds appear.

DRAINAGE. We have already seen (p. 58) that ordinary cultivated plants will not thrive in a waterlogged soil. Wherever there is an excess of water it must be removed before satisfactory results can be obtained. Fig. 47 shows a field of wheat in May where the crop is all but killed and only certain weeds survive on a patch of undrained land that lay wet all the winter. Draining land is difficult and somewhat expensive; trenches are first cut to a proper depth, and drain pipes are laid on the bottom, taking care that there is a gentle slope all the way to the ditch. The rain soaks into the soil and gets into the pipes, for they are not joined together like gas or water pipes, but left with little spaces in between; it then runs out into the ditch.

Fig. 47. A badly drained wheat field.

Usually only clay soils or peats need drainage, but occasionally sandy soils do also (see pp. 22, 92). A great deal of drainage was carried out in England between 1840 and 1860, and it led to a marked improvement in agriculture and in country life generally; it almost put an end to ague and other troubles that used to afflict the villagers. Then for many years very little drainage was done, but it has been started again, and draining machinery, mole ploughs, etc. can be seen working in many places.

LIMING. The addition of chalk or lime to soil was found in chapter III to make it less sticky and less impervious to air and water. Chalk or lime does more than this. It puts out of action certain

Fig. 48. Lime benefits an acid soil: Tunstall, East Suffolk. The background has been limed and the sugar beets are growing, the foreground has had no lime and the sugar beets have failed. But the rye is hardly affected.

injurious substances or acids that may be formed, and this makes the conditions more favourable for plants and for the useful micro-organisms; farmers and gardeners express this by saying that it 'sweetens the soil'. As we have seen on p. 68, many soils in England are improved by adding lime or chalk (Fig. 48).

There are considerable areas in the south-eastern and eastern counties where the soil is very chalky; here you find a wonderfully rich assortment of flowers and shrubs. Where there is too much chalk the soil is not fertile; some plants lose their colour ('chlorosis'), and chalk lets water through too easily (p. 21): but this makes it admirable for residential purposes.

Some soils lying on chalk or limestone are lacking in lime. This seems odd, but it often happens under high rainfall or where the soil has been formed, not from the chalk, but from other material.

Lime really differs from chalk, but changes into it so quickly in the soil that the action of both is almost, though not quite, the same.

ADDING PLANT FOOD TO THE SOIL. We have already seen (p. 40) that leaves and stems when put into the soil give rise to plant food. Chemists have discovered what these foods are, and how they can be prepared on a large scale: they are now sold to farmers and gardeners and called FERTILIZERS. There are three kinds; nitrogenous, phosphatic, and potassic. They are very strong and must be used only in small quantities or they may do harm: half an ounce to an ounce per square yard is plenty for nitrogenous fertilizers, about double these quantities for phosphatic; and rather less than double for the potassic or potash fertilizers. It is often most convenient to use a mixture of all three, but you should make an experiment in the school garden with the separate substances to see how they act. It should be begun in early spring.

Measure out six plots each of 1 square yard. Prepare all equally well for sowing; leave one unmanured, to another add sulphate of ammonia or nitrate of soda (nitrogenous) at the rate of 1 oz. per square yard, to the third add 2 oz. superphosphate (phosphatic), to the fourth add $1\frac{1}{2}$ oz. muriate of potash (potassic), and to the fifth add a mixture of all three in the same quantities as on the other plots, i.e. $4\frac{1}{2}$ oz. in all. The fertilizer must be put on very evenly, and is best watered in. To the sixth add farmyard manure as a good gardener would do it. Then sow some leafy crop such as barley or mustard (but not peas or beans) over the whole area and watch carefully the results, measuring from time to time the average height of the plants on the different plots. When the plants have reached full growth cut them and weigh them, then leave them to dry in a warm place and weigh them again.

The results depend on the amount of plant food naturally present in the soil, but generally the nitrogenous fertilizer gives the plants a deeper green colour, hastens their growth, and gives an extra weight at the end. The effects of the phosphate and potassic fertilizer depend on the soil. Usually the mixture of all three gives the best results.

This simple experiment shows the chief effects, but it is not accurate enough to help farmers. To do it properly each treatment must be repeated several times but not in any regular order: the treatments are all jumbled up, and the arrangement is left purely to chance; it is determined by drawing numbers out of a bag, by drawing cards, or some similar device.

Experiments of this kind are carried out in many parts of the country, and if there is one in progress near enough you should see it. But you must not be disappointed if you do not fully understand it for it may be very complicated: the experimenter knows that he cannot avoid experimental errors (p. 39) so he tries to find out how big they are, and this jumbled repetition of the plots is his surest method. If the differences in weight are well above the experimental error they can be taken as being produced by the fertilizer; if they are not, they may quite well be due to something else and not to the fertilizer at all. It is by means of such experiments that scientists advise farmers how they should manure their crops.

In the Appendix particulars are given of an experiment like this that has been made in many school gardens. It has been used to test not only the effects of fertilizers but questions much debated by the village gardening experts, for example, what varieties to use, whether direct seeding or transplanting is better, whether spacing should be wide or narrow. The experiment requires care and attention, but many have done it well and you may like to try it.

ORGANIC MATTER: HUMUS. You remember that the top soil is much better than the subsoil for the growth of plants (p. 35) and that it contains more of the dark-coloured combustible material derived from plant residues. This material is called humus and it is one of the chief causes of the difference between the top and the subsoil. You should now go through your notes on the previous four chapters and collect all the information you can about this humus. Then you should compare a soil rich in humus with one that contains less: the fairest comparison is an old well-manured garden soil with that from an adjoining arable field. We had a nice example at Harpenden. The soil was naturally heavy, but the garden soil was more easy to cultivate; it gave a better tilth, more easily formed crumbs, and produced bigger and better plants than the field soil. This we expected because we knew that the black material furnished plant food. But another difference was found. The garden soil always contained more moisture than the field soil:

	6 April	6 May	6 July	28 October
Moisture in dark soil rich in humus, per cent.	20·0	18·0	20·7	23·3
Moisture in lighter soil poor in humus, per cent.	13·1	11·9	12·0	17·5

This, of course, is a help to the plant, but not as much as it looks, because the humus sticks to some of the water and will not give it all up to the plant.

An important way of improving the fertility of a soil is to increase the amount of humus it contains. This has to be done regularly because, as we have seen (chapter VII), the organic matter is continuously being consumed by the dwellers in the soil. So every four or five years farmers plough in farmyard manure or a grass ley when the animals have finished with it, and gardeners almost every year dig in farmyard manure or compost. This is so important that every garden should have its compost heap made of all available vegetable matter (unless it is diseased, or very tenacious of life like bindweed, couch grass, docks etc.), household waste, and animal manure of any sort.

Fig. 49. Ploughed up hill grass land for arable crops: Wallog, Cardiganshire.

There is sometimes much argument whether organic manures or fertilizers are best. This is like arguing whether bread or butter is the better. The two things are so different that they cannot be compared, and the proper answer is that both are needed. But, and this is very important, fertilizers must be used with knowledge and understanding or they may do harm, and that is where experiments like those on p. 84 are so necessary.

Old grass lands gradually accumulate humus but not indefinitely;

beyond a certain point the losses caused by the dwellers in the soil about equal the gains so that the amount does not increase; when the grass land is ploughed some of the humus is decomposed and yields a certain amount of plant food. During the wars of 1914–18 and 1939–45 much grass land was ploughed and sown to wheat and other crops. A great deal of it was less productive than some people had expected, showing that humus alone would not give full crops; when, however, lime and fertilizers were added the results were much better. It is known that all these are needed for good plant growth. Fig. 49 shows a hill farm in Cardigan, formerly all grass, but some of it was ploughed during the war and when properly fertilized produced much more food than it had done before.

SUMMARY. The various things we have learnt in this chapter are:

Autumn and winter cultivation are needed to loosen the soil so that rain can soak in and not lie about in pools, and also to facilitate working in spring.

The soil has to be broken down very finely and made rather dry for a seed-bed. The seed has to be pressed or rolled in and then left entirely alone.

As soon as the little plants are up the soil must be hoed to break up any crust that has formed, to keep the soil cool in hot weather, and to destroy weeds, which successfully compete against almost any cultivated plants.

Drainage may be necessary to remove excess of water.

Liming or chalking the soil is beneficial, not only because of the improvements mentioned in chapter III, but also because certain injurious substances are thereby put out of action. There are, however, some plants that will not tolerate lime.

Plant foods can be added to the soil in the form of fertilizers and if properly used they give good increases in plant growth. But if wrongly used they may do harm. They do not supply organic matter. This must be given as farmyard manure, compost, etc.

CHAPTER X

THE SOIL AND THE COUNTRYSIDE

In this chapter we want to put together much of what we have learned about the different kinds of soil, so that as we go about the country we may know what to look for on a clay soil, a sandy soil, and so on.

We have seen that clay holds water and is very wet and sticky in winter, while in summer it becomes hard and dry, and is liable to crack badly. Clay soils are therefore hard to dig and expensive to cultivate: the farmer calls them HEAVY and usually prefers to put them into grass because once the grass is up it can be left without much attention, though of course it does not then make much growth. But in the days when we grew our own wheat, before we imported it from Canada, Australia, and other countries, this clay land was widely cultivated for wheat and beans. So long as wheat was 60*s*. to 100*s*. a quarter it was a very profitable crop, but, when in the 1880's and 1890's its price fell heavily the land either went out of cultivation or it was changed to grass land and used for cattle grazing. Great was the distress that followed; some districts indeed were years in recovering. But new methods came in: the land near London was used for dairy farming, and where suitable for potato growing, while elsewhere it was improved for grazing; the clay districts, completely changed, became more prosperous again. Many of the fields still show the ridges or 'lands' in which, when they grew wheat, they were laid up to let the water run away, and many of them keep their old names, but these are the only relics of the old days. The land is not, and never was, agriculturally very valuable. Some of the old commons still remain, rough with mole hills and ant hills. Bracken, gorse, rushes, thistles and brambles grow there, and you may find many fine blackberries in September. The coarse *Aira* grass is found with its leaves as rough as files. The villages are often built round greens which serve as the village playground, where the boys and young men now play cricket and football, and their forefathers practised archery, played quoits and other games. On a few village greens the ancient stocks, in which offenders were placed, can still be seen.

There are numerous woods and plantations containing much oak. Some of the woods are very ancient and may even form part of the primeval forests that once largely covered England. Epping Forest in Essex, the Forest of Blean, the King's Wood in Kent, and others, have probably never been cultivated land. In the days when ships were made of oak these woods and hedges were very valuable, but now they are less used as sources of timber. For long they were valued for quite another reason: they afforded shelter for foxes and for game birds. The clay districts are and always have been famous for fox hunting; the Pytchley, Quorn, Belvoir, and other celebrated packs have their homes in the broad clay, grassy vales of the Midlands. The vale of Blackmoor and other clay regions are equally famous. The plantations and hedgerows are fine places for primroses and foxgloves, while in the pastures, and especially the poor pastures, are found the ox-eyed daisy and quaking grass, that make such fine nosegays, as well as that sure sign of poverty, the yellow rattle. But many of these poor pastures have been improved by draining, liming, and the use of suitable manures. The roads used to be very bad (see p. 24) and old people can still remember when they would lie wet for weeks in winter, especially where the hedges were high. Numerous brick and tile yards may be found, some still working, some worked out.

A sandy soil is in so many ways the opposite of a clay soil that we shall expect to find corresponding differences in the look of the country. A sandy soil does not hold water: it may get water up from the subsoil to supply the plant if the water table is near enough to the surface (see p. 55), or, if the soil happens to lie in a basin of clay, it may even be very wet: otherwise it may be too dry for ordinary plants. We may therefore look out for two sorts of sand country, the one cultivated because there is enough water for the crops, and the other not cultivated because the water is lacking. These can readily be found.

We will study the cultivated sands first. As sand is not good plant food (p. 35) these soils want a lot of manure and often lime as well; so they are not good for ordinary farming of the old-fashioned sort. But they are very easy to cultivate—for which reason they are called LIGHT soils—and can be dug at any time; seeds can be sown early, and early crops can be got. Consequently these soils are very useful for producing special crops like fruit, all sorts of vegetables, carrots, parsnips, broccoli, peas, for special dairy farming, and so on. Fig. 50 is a view of a highly cultivated sandy region in Kent showing strawberries in the foreground, bush fruit behind, and fruit trees behind that again.

The uncultivated sands are sometimes not really so very different, and some of them have in recent years been made to grow crops. But they always require special treatment and therefore in the old days they were generally left alone. Our ancestors disliked them very much; 'villainous, rascally heaths' Cobbett always called them. There were practically no villages and few cottages, because the land was too barren to produce enough food; the few dwellers on the heath, or the 'heathen', were so ignorant and benighted that the name came to stand generally for all such people and has remained in our language long after its original meaning was lost. As there were so few inhabitants the heaths used to be great places for robbers, highwaymen, and evildoers generally; Gad's Hill on the Watling Street between Rochester and Gravesend, Finchley Common,

Hounslow Heath and others equally dreaded by travellers of the seventeenth and eighteenth centuries, were barren sandy tracts. But fashions change: these heaths are now fashionable residential places; they are in many places dotted with red-bricked and red-tiled villas, and have lost their ancient character.

Fig. 50. Highly cultivated sandy soil, Charing Heath, Kent. Strawberries in foreground, behind them raspberries and currants, in background cherries and apples.

The heaths are not everywhere dry; there are numerous clay basins where the sand lies wet, where peat forms (see p. 22), and where marsh plants like the bog asphodel, sundew, or cotton-grass can be found. In walking over a heath you soon learn to find these wet places by the colour of the grass and the absence of heather. In some places there is a good deal of wood, especially pines, larches, and silver birches: all these are very common on the Surrey sands; willows also grow in the damp places. Fig. 51 shows a Kent heath—Hothfield—with heather, gorse and bracken; with pine woods in the distance and frequent bare patches of sand. Much of the New Forest is on the sand, as also is Bournemouth, famous for its fine

Fig. 51. A Kent heath, Hothfield.

pine woods. But elsewhere there are no trees: in the old days the peasants used to burn the turf or the gorse after they had fired it to destroy the prickles, and so their cottages had huge fireplaces: instead of fences they used walls made of turf. Such are the Dorchester heaths described by Thomas Hardy in his novels. Other sands, however, are covered with grass and not with heather, and many of these have a special value for golf links, especially some of the dry invigorating sands by the seaside. The famous links at St Andrews, and at Littlestone, are examples.

In between the fertile and the barren sands come a number that are cultivated without being very good. They are much like the others, carrying a vegetation that is usually of the narrow-leaved type (p. 60), and not very dense. On the road sides you see broom, heather, heath, harebells, along with gorse and bracken with milk-wort nestling underneath: crested dog's tail and sheep's fescue are common grasses, while spurrey, knotweed, corn marigold, are a few of the numerous weeds in the arable fields. Gardens are easily dug, but it is best to put into them only those plants that, like the native vegetation, can withstand drought and do not need lime; vegetable gardens must be well manured and well limed.

Even in the old days it was always easy to travel in a sand country because the roads dried very quickly after rain, although they might be dusty in summer. Sometimes the lanes are sunk rather deeply in the soft sand, forming very pretty banks on either side.

Loams, as we have seen (p. 1), lie in between sands and clays: they are neither very wet nor very dry: not too heavy nor yet too light: they are very well suited to our ordinary farm crops, and they form by far the best soils for general farming; wheat, oats, barley, sheep, cattle, milk, fruit and vegetables can all be produced: indeed the farmer on a good loam is in the fortunate position of being able to produce almost anything he finds most profitable. In a loam district that does not lie too high the land is generally all taken up, even the roads were narrow till recently and there are few commons. The hedges are straight and cut short, the farm houses and buildings are well kept, and there has nearly always been a general air of prosperity all round. Good elms grow and almost any tree that is planted will succeed. Loams shade off on one side into sand; the very fertile sands already described might quite truly be called sandy loams.

On the other side they shade off into clays; the heavy loams are splendid wheat soils and grow sugar beet well. Cultivation used to be difficult but tractors have now made it much easier. They form pleasant, undulating country, nicely wooded, and dotted over with thatched cottages; the fields are less wet, and the roads in old days were rather better than on the clays. When properly managed they make excellent grass land.

Chalky soils stand out quite sharply from all others: their white colour, their lime kilns now often disused, their noble beech trees, and, above all, the great variety of flowering plants enable the traveller at once to know that he is on the chalk. Many plants like chalk and these may be found in abundance, but some, such as foxgloves, heather, broom or rhododendrons cannot tolerate it, and so they will not grow.

Chalk, like sand, is dry, and the roads could always be traversed very soon after rain. The by-roads and lanes are often narrow, winding, and worn deep especially at the foot of the hills, so that the banks get a fair amount of moisture and carry a dense vegetation. Among the profusion of flowers you can find scabious, the bedstraws, vetches, ragwort, figwort, and many a plant rare in other places, like the wild orchids, while the cornfields are often yellow with charlock. In the hedgerows are hazels, guelder roses, maples, dogwood, all entwined with long trails of bryony and traveller's joy. In the autumn the traveller's joy produces the long, hairy tufts that have earned for it the name of old man's beard, while the guelder roses bear clusters of red berries. The great variety of flowers attracts a corresponding variety of butterflies, moths and other insects; there are also numbers of birds and rabbits—indeed a chalk country teems with life in spite of the bare look of the Downs. The roads running at the foot of the chalk Downs and connecting the villages and farmhouses built there for the good water supply, are particularly rich in plants because they sometimes cut into the chalk and sometimes into the neighbouring clay, sand or rock. Now and then a spring bursts out and a little stream takes its rise: if you follow it you will generally find watercress cultivated somewhere.

Besides the beech trees you also find ash, sycamore, maples, and, in the church yards, some venerable yews. Usually the chalk districts were inhabited very early: they are dry and healthy, the land can be cultivated and the heights command extensive views over the country, so that approaching enemies could easily be seen. On the chalk downs and plains are found many remains of tribes that lived there in the remote ages of the past, whose very names are now lost. The oldest of them go back some 4500 years. Strange weapons and ornaments are sometimes dug up in the places where they lived and worked; the barrows can be seen in which they were buried, and the temples in which they worshipped; Stonehenge itself, the best known of all these, lies on the chalk. Some of their roads are still in use to this day, the Icknield Way (the way of the Iceni, a Belgic tribe), the Pilgrim's Way of the southern counties and others.

Even the present villages date back to very ancient times, and the churches are often seven or eight hundred or more years old.

On the more level ground where cultivation is possible the fields are very big and without hedges, they are admirably suited to tractors and large implements. Many of the farms cover several hundred acres, and are highly mechanized. In the old days great flocks of sheep were kept and the whole scheme of farming was based on their requirements. Arable crops were grown on which they were 'folded'. The dry conditions of the chalk suited them admirably. Shepherds were then very important people; you should read a book W. H. Hudson wrote about them: it is called *A Shepherd's Life*. But nowadays dairy cattle are more profitable than sheep and so have largely taken their place, and those that are left are managed differently.

The limestone country of the Cotswold region is like the chalk country in many ways, for chalk and limestone are both made of the same chemical substance, calcium carbonate.

Limestone, however, is harder than chalk, and so is much used for building walls and houses, even for roofing them. So the limestone country is a stone-built country: brick is out of place and should not be used. The Cotswold region forms the largest area of limestone in the south of England; like the chalk region it used to be famous for its sheep and had been for many years; from the fourteenth century onwards wool brought much wealth to the countryside. Some of the limestone can be carved, and the village craftsmen learned to do this well: in many of the Cotswold villages there are very beautiful churches, and some of the little towns like Burford, Chipping Camden, and others are extremely attractive. Unfortunately, much of the modern building is very poor and looks dreadful compared with the old. As in the chalk region, sheep have been largely displaced by dairy cattle.

Another kind of limestone occurs in the north of England, also very hard and hence used for making walls and building houses: this too is a stone country. But the stone is not suitable for carving and so neither villages nor churches have the beauty of the Cotswolds. Also the soils lying on the limestones are quite different from the chalk soils; often they have come from another district, having been carried by the glaciers in the days when England was buried under the ice. Many of them are poor, and need both fertilizers and lime to make them productive. In winter you often see in the fields heaps of lime that farmers have carted out to spread (Fig. 52).

The black soil of the fen districts and elsewhere is widely different from any of the preceding. It contains, as its colour shows, a large quantity of combustible material (chapter v), which has a great power of holding water. These fens are therefore very wet; until they were drained they were desolate wastes: you may read in

Kingsley's *Hereward the Wake* what they used to be like in old days, and even as late as 1662 Dugdale writes that here 'no element is good. The air cloudy, gross and full of rotten harrs;* water putrid and muddy, yea, full of loathsome vermin; the earth spongy and boggy; and the fire noisome by the stink of smoking hassocks.'†

Fig. 52. Heaps of lime waiting to be spread near Matlock, Derbyshire.

But during the Stuart period wide ditches or drains were dug, into which the water could flow and be pumped into rivers. This reclamation has been continued to the present time, and the black soils as well as the others in the Fen districts are now very productive. The rain-

* Harr is an old word meaning sea-fog.

† Hassock is the name given to coarse grass which forms part of the turf burnt in the cottages.

fall is low and they grow splendid crops of wheat, potatoes and sugar beet, besides special crops like celery and mustard—this explains why mustard comes from Norwich.

But the soils are very light and like sands (p. 18) can blow away. This is what happened in the field shown in Fig. 53: it was sown,

Fig. 53. Wind erosion on a recently reclaimed light 'blowing' fen soil: the first crop was blown away and a second is being sown (Wissington Fen, Norfolk).

but while the plants were small a great wind arose and carried them off along with the top layer of soil. The farmer is now re-sowing and hopes for better luck next time.

Fig. 54 shows the change in Borth bog, Cardigan, after drainage and reclamation. Formerly it was a wet waste; now this part has been sown with grass and provides food for cattle. Some can become arable land, but owing to the high rainfall grass is always the safest crop.

We have seen that a change in the soil produces a change in the plants that grow on it. The FLORA (i.e. the collection of plants) of

Fig. 54. Reclaimed part of Borth bog, Cardigan.

a clay soil is quite different from that of a sandy soil, and both are different from that of a chalk or of a fen soil. In like manner draining a meadow or manuring it alters its flora: some of the plants disappear and new ones come in. Even an operation like mowing a lawn, if carried on sufficiently regularly, causes a change. In all these cases the plants favoured by the new conditions are enabled to grow rather better than those that are less favoured; thus in the regularly mown lawn the short growing grasses have an advantage over those like brome that grow taller, and so crowd them out. When land is drained those plants that like a great quantity of water no longer do quite so well as before, while those that cannot put up with much water have a better chance. In the natural state there is a great deal of competition among plants, and only those survive that are adapted to their surroundings. You should remember this on your rambles, and when you see a plant growing wild you should think of it as one that has succeeded in the competition and try to find out why it has been enabled to do so.

CHAPTER XI

HOW SOIL HAS BEEN MADE

APPARATUS REQUIRED. The apparatus in Fig. 58. The under surface of the lips of the beakers should be vaselined to prevent the water trickling down the sides.

Rocks look very solid but in time they get broken into fragments by the weather and especially by the action of freezing water (p. 72). If the rock is fairly level the fragments may stay there. At the top of a quarry you can notice that the solid rock forms a kind of a floor, above that it lies in great splintered lumps, above are smaller pieces and on top is the soil covered with vegetation (Fig. 55). All have been formed from the rock, and the process is called WEATHERING. At each stage some small particles have broken off but they have not all remained where they were formed; in course of time earthworms have brought them to the surface, making a layer on which plants can grow: these supply the organic matter necessary to produce a top soil. Great areas of our soil have been made in this way.

But more often the soil does not stay where it was formed; it is carried away by water. This is most easily done from a hillside, but it also happens on gently sloping ground. You can see that from the mud in the streams. Often too the rock is not level: for example, on the face of a cliff, the steep side of a mountain or of a disused quarry, and there the rock fragments fall and form a slope up which you can scramble: if left long enough the fragments may splinter still further, and mosses, lichens and plants may start growing: in course of time a soil develops. But here too, the small particles keep getting washed away. If there is a stream, river, or sea at the foot of the cliff the fragments may be carried away as fast as they roll down:

the differences shown in Figs. 56 and 57 between a cliff at the seaside and a cliff inland arise simply in this way. In inland districts great valleys are in course of time carved out, and at the seaside large areas of land have been washed away.

Fig. 55. Quarry section showing soil formation. Yellow Guiting stone quarry, about 1 m. N.E. of Temple Guiting church, Gloucester.

What becomes of the fragments thus carried away by the water? The best way of answering the question would be to explore one of these mountain streams and follow it to the sea, but we can learn a good deal by a few experiments that can be made in the classroom. We want to make a model stream and see what happens to little fragments of soil that fall into it.

Fix up the apparatus shown in Fig. 58. The small beaker *A* is to represent the narrow mountain stream, the larger one *B* stands for the wide river, and the glass jar *C* for the mouth of the river or the sea. Run water through them; notice that it runs quickly through *A*, slowly through *B*, and still more slowly through *C*: we want it to do this, because the stream flows quickly and the river slowly.

Fig. 56. Cliffs at the seaside, Manorbier. The broken material is washed away.

Now put some soil into *A*. At once the soil is stirred up, the water becomes muddy, and the muddy liquid flows into *B*. But very soon a change sets in, the liquid in *A* becomes clear, and only the grit and stones are left in the bottom: all the mud—the clay and the silt—is washed into *B*. There it stops for a long time, and some of it will never wash out. The liquid flowing into *C* is clearer than that flowing into *B*. If you keep on putting fresh portions of soil into *A* you can keep *B* always muddy, although *A* is usually clear. At the end of the experiment look at the sediment in each beaker: in *A* it is clear and gritty, in *B* it is muddy. If you can get hold of some sea water put some of the liquid from *C* into it: very soon this liquid clears and a deposit falls to the bottom, the sea water thus acting like the lime water on p. 16.

Fig. 57. Inland cliffs, Salisbury Craigs, Edinburgh: the broken material remains as a slope.

The experiment shows us that the fine material washed away by a quickly flowing stream is partly deposited when the river becomes wider and the current slower, and a good deal more is deposited by the action of the salt water when the river flows into the sea. The rock that crumbles away inland is spread out on the bed of the river or at its mouth.

The river Stour at Wye showed all these things so clearly that I will describe it; you must then compare it with a river that you know, and see how far the same features occur. At the bridge the stream was shallow and flowed quickly: the bottom was gritty and pebbly, free from mud, and formed a safe place for paddling. Before the bridge was built there had been a ford here. But farther away, either up or down, the stream was deeper and wider, flowed more slowly, had a muddy bottom, and so was not good for paddling. At one place about a mile away some one had widened out the river to

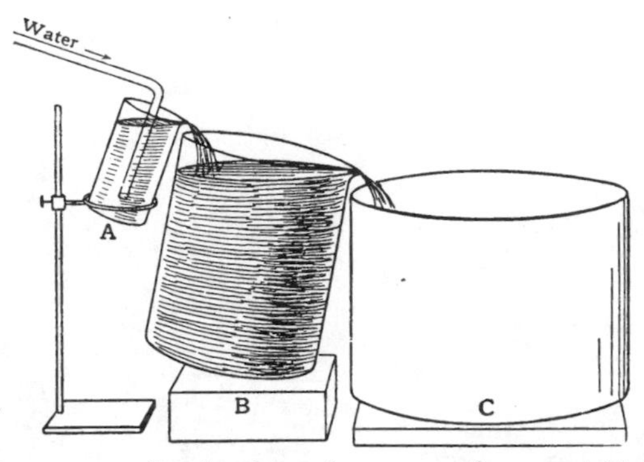

Fig. 58. Model of a stream. In *A*, where the stream flows quickly, the water is clear and the sediment free from mud. In *B*, where it flows slowly, the water is turbid and the sediment muddy. In *C*, the sediment contains much of the clay.

form a lake, but this made the stream flow so slowly (as it was now so much wider) that the silt and clay deposited and the lake became silted up, i.e. it became so shallow that it was little more than a lake of mud. The same facts were brought out at the bend of the river. On its convex side, Fig. 59, the water has rather farther to go in getting round the bend than on its concave side *B*; it therefore flows more quickly, and carries away the soil of the bank and mud from the bottom. But on its concave side where it flows more slowly it deposits material. There is at the bend a marked difference in depth at the two sides. On its convex side the stream is rapid and deep, and scours away the bank; on its concave side it is slower, shallower, and tends to become silted up. Thus the bend becomes more and more pronounced unless the bank round *A* is protected (the other bank of course needs no protection) and the whole river winds about just as you see in Fig. 60, and is perpetually changing its course,

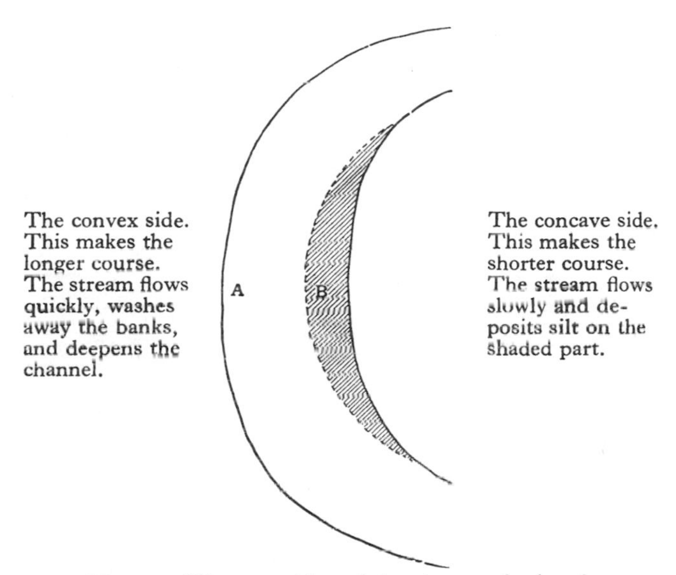

Fig. 59. The two sides of the river at the bend.

carrying away material from one place, mixing it up with material washed from somewhere else, and then deposits it at a bend or in a pool where it first becomes a mud flat and then dry land. Some, however, is carried out to sea. We need not follow the Stour to the sea; reference to an atlas will show what happens to other rivers. Some of the clay and silt they carry down is deposited at their mouths, and becomes a bar, gives rise to shoals and banks, or forms a delta. The rest is carried away and deposited on the floor of the sea. Material washed away by the sea from the coast is either deposited on other parts of the coast, or is carried out and laid on the floor of the sea. Thus a thick deposit is accumulating, and if the sea were to become dry this deposit would be soil. This has actually happened in past ages. The land we live on, now dry land, has had a most wonderful history; it has more than once lain at the bottom of the sea and has been covered with a thick layer of sediment carried from

Fig. 60. The winding river—the Stour at Wye.

other places. Then the sea became dry land and the sediment, which had become pressed into rock, formed new soil, but this at once began to get washed away by streams and rivers into new seas, and gave rise to new sediments on the floor of these seas. And so the rock particles have for untold ages been going this perpetual round: they become soil; they are carried away by the rivers, in time they reach the sea; they lie at the bottom of the sea while the sediment gradually piles up: then the sea becomes dry land and the process starts all over again. The eating away of the land by water is still going on: it is estimated that the whole of the Thames valley is being lowered at the rate of about one inch in eight hundred years. This seems very slow, but eight hundred years is only a short time in geology, the science that deals with these changes.

Water does more than merely push the rock particles along. It dissolves some of them, and in this way helps to break up the rock. Spring water always contains dissolved matter, derived from the rocks, some of which comes out as 'fur' in the kettles when the water is boiled.

Rocks are also broken up by other agents. There is nearly always some lichen living on the rock, and if you peel it off you can see that it has eaten some of the rock. When the lichen dies it may change into food for other plants.

We have learnt these things about soil formation. First of all the rocks break up into fragments through the splitting action of freezing water, the dissolving action of liquid water, and other causes. This process goes on till the fragments are very small like soil particles. Earthworms sort out these small particles bringing them to the surface where they form a layer on which plants begin to grow, and as these die and decay they give rise to the black humus that we have seen is so valuable a part of the soil (p. 86). This is how very many of our soils have been made. But the action of water does not stop at breaking the rock up into soil; it goes further and carries the particles away to the lower parts of the river bed, or to the estuary, to form a delta, and mud flats that may be reclaimed, like Romney Marsh in England and many parts of Holland have been. Many of our present soils have been formed in this way. Finally the particles may be carried right away to sea and spread out on the bottom to lie there for many ages, but they may become dry land again and once more be soil.

This process is helped by building a wall for protection against the high tides: a good deal of land has been gained in this way round the coast of England (Figs. 61 and 68). These changes are going on always and everywhere. Our hills and mountains used to be much

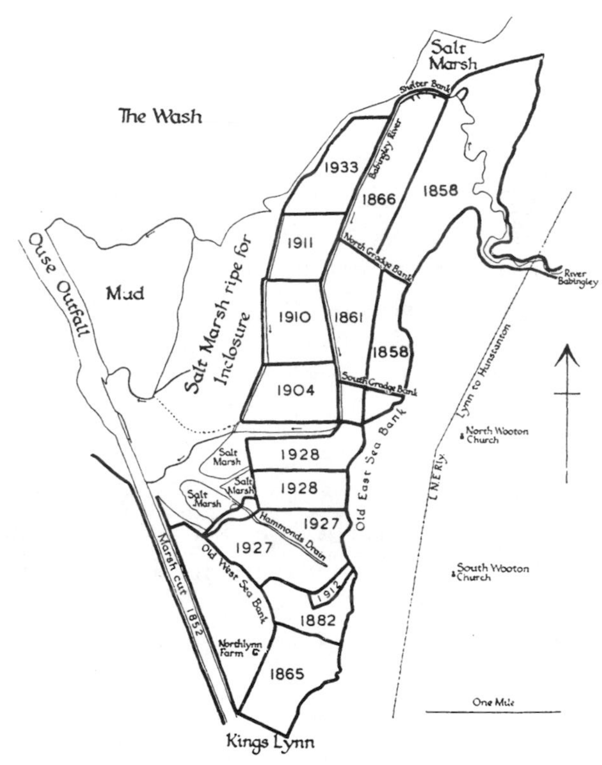

Fig. 61. Reclamation of land round the Wash from the sea. Map showing the enclosures and the dates when they were made. The old sea bank is now far inland.

higher than they are now, and they keep on shrinking: nature is never at rest; water and wind (pp. 18, 99) are continually wearing away our land. This is called EROSION: it is taking place all over the world. When the land is covered with forest or grass erosion may be extremely slow, but when the forest is cut down and the grass ploughed up, it may become very rapid: in some countries highly destructive. Fortunately in Great Britain we suffer but little because we take care, but Fig. 62 shows what may happen on ploughed land, and Fig. 12 shows what wind can do. The establishment of a vegetative cover and, where necessary, of belts of trees as shelter from the wind, protects the soil and reduces the amount of erosion.

Fig. 62. Soil erosion caused by rain storm, near Caistor, Lincolnshire.

Look at Fig. 63. It is a picture of the hard stone country in Northumberland. Notice the sharp outlines and the stone walls. The fields lie too high for arable cultivation and are therefore left in grass so the farming must depend on sheep and cattle. There are trees in sheltered places but they are rounded; the winds do not encourage tall growth. The stone walls give the sheep some shelter from the snow in the cold winters but only very hardy animals thrive here.

Fig. 63. Northumberland hill farm: mainly sheep and cattle.

Fig. 64. Dunster: fertile valley, the soil washed from the upland.

Now turn to Fig. 64. That is also in a stone country, but it is in the warmer, moister and more genial climate of Somerset, so the trees can grow tall and well. Cultivation is pushed up the slopes, and even on to the tops.

Fig. 65 shows a fertile undulating region, lower lying and less windswept, therefore trees can grow on the hill tops; much of the

Fig. 65. Rolling country rich in lime: Chilmark, near Salisbury.

land is arable and the fields are large to enable big implements to work; the good dairy herd in the foreground, the large farm-house and well-built cottages testify to the productiveness of the land.

But we must now return to the River Stour; at the Wye School we learned something more about it. Why did it flow quickly at the bridge and slowly elsewhere? We knew that the soil round the bridge was gravelly, whilst up and down the stream it was clayey. The river had not been able to make so wide or so deep a bed through the gravel as it had through the clay, and it could therefore be forded here. We knew also that there was a gravel pit at the next village on the river, where also there was a bridge and had been

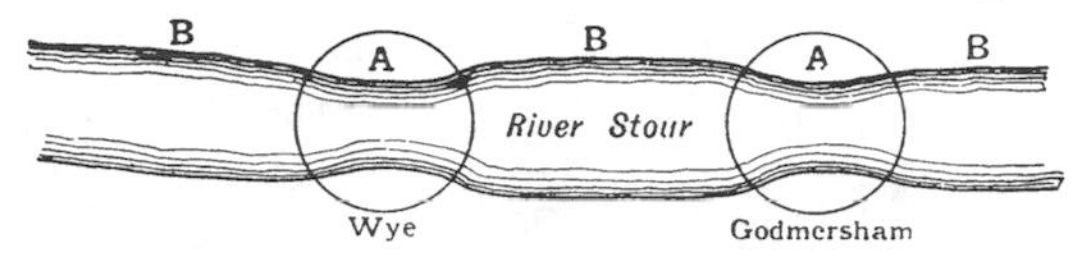

Fig. 66. Sketch map showing why Godmersham and Wye arose where they did on the Stour. At *A*, the gravel patch, the river has a hard bed and can be forded. A village therefore grew up here. At *B*, the clay part, the river has a soft bed and cannot be forded. The land is wet in winter, and the banks of the stream may be washed away. It is therefore not a good site for a village.

Fig. 67. Ford at Coldharbour, near Harpenden.

a ford, and so we were able to make a rough map like Fig. 66, showing that fords had occurred at the gravel patches, but not at the clay places. Now it was obvious that an inn, a blacksmith's forge, and a few shops and cottages would soon spring up round the ford, especially as the gravel patch was better to live on than the clay round about, and so we readily understood why our village had been built where it was and not a mile up or down the stream. Almost any river will show the same things: on the Lea near Harpenden we found the river flowed quickly at the ford (Fig. 67), where there was a hard, stony bottom and little mud: whilst above and below the ford

Fig. 68. Land reclaimed from the Wash.

the bottom was muddy and the stream flowed more slowly. At the ford there is as usual a small village. The Thames furnishes other examples: below Oxford there are numerous rocky or gravelly patches where fords were possible, and where villages therefore grew up. Above Oxford, however, the possibilities of fording were fewer, because the soil is clay and there is less rock; the roads and therefore the villages grew up away from the river.

CHAPTER XII

A SURVEY OF YOUR DISTRICT

If you have faithfully made the experiments and observations indicated in the preceding chapters you will have accumulated much interesting information about the district in which you live. You should now try to collect this all together, so as to leave a record of your district for those who will come after you and who may find many changes. For this you will need the quarter-sheets of the Ordnance Survey 6-inch maps. Each of these covers an area measuring 3 by 2 miles, but it may happen that your district spreads over two or even four of them. You should form into groups, each of which would survey one part of the area. Each group should make a tracing of the part of the map dealing with its area; roads, field boundaries, woods, commons and any other items of agricultural interest should be included, but you need not include details of towns or villages, except allotments. Number each field. Permission should be sought for each group to visit every part of its section.

Then, armed with your permit, note-book, and tracing, visit every field in your part of the area and note the use to which it is put. Use the letters given below, write them clearly in pencil, then when you return to school or home write them in ink and paint each field with the proper colour-marking.

The letters and colours should be:

	Letter	*Colour*
Arable land: crops, fallow, rotation grass (up to 3 years old)	A	Brown
Market gardens (vegetables, small fruits)	A (MG)	Brown
Grass land: four or more years old	M	Light green
Heathland: moorland, commons, rough hill pasture, grazed wet land	H	Yellow
Gardens: allotments, orchards, nurseries	G	Purple
Forest and woodland	F	Dark green
Unproductive agriculturally: buildings, yards, mines, cemeteries	W	Red
Ponds, lakes, reservoirs, and anything containing water.	P	Blue

Enter in a note-book the numbers you have given to the fields on your tracing. Against each number write down the name of the field, what crop it was carrying, and whether the soil is good, moderate or poor; if it is poor, put down why: too dry, too wet, too acid, 'hungry', or whatever the reason may be. If you can test the soil with the Acid Indicator do so. Note also any springs.

The arable (A) and the grass land (M) will not usually give you much trouble, except that you may not always be able to distinguish rotation from permanent grass if it is in a field by itself. Where the

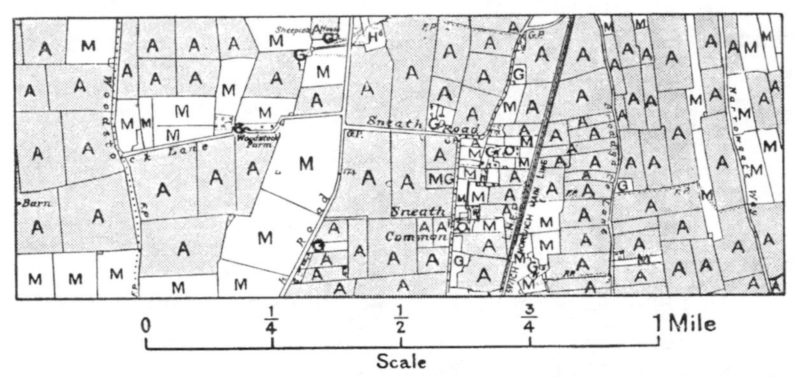

Fig. 69. Working sheet, Land Utilization Survey, Sneath, south Norfolk. A, arable land; G, gardens; M, grass land; O, orchard; MG, market garden.

stubbles of the previous corn crop are visible you can be sure that it is rotation grass, and if there is much red clover you can be nearly sure; if, however, there are many thistles, nettles, rushes or much wild white clover it is almost certainly permanent grass; in case of doubt you should ask the farmer or one of the workers. You should distinguish between good, moderate and poor permanent grass. In chalk or limestone regions the open down land covered with short herbage regularly grazed by farm animals comes into the M group, particularly if it is ever manured, but it is in the H group if it is so thickly set with bushes as to make it only of little value for grazing. Mountain and high hill stretches of rough grazing are also H. Private parks are generally M, except definite plantations which are F.

Heathland (H) is distinguished from grass land (M) by its large amount of vegetation unsuited for close grazing, except perhaps when it is very young, for example, bracken, heather, gorse, bramble, thorn and other bushes: it can be used only for rough grazing, usually it forms big open stretches not divided into fields; unfenced except on the outside boundary. No farming operations are done on it, though if it is very wet it may be drained by means of deep ditches.

Gardens (G) include all house gardens big enough to grow vegetables and flowers. Backyards, however, are not counted. Orchards are in some districts grazed and should then be marked G (M) because they are both orchard and grass; if vegetables or other crops are grown in them they are both orchard and market garden and so become G (A).

Unproductive land (W) differs from heathland in being permanently impossible for agricultural use; for example, land covered with buildings, concrete, waste heaps, unless covered with vegetation, when they become H. In the note-book you should record why you have classed the land W.

All notes should be dated and signed. The records on the tracings should then be transferred neatly to the map: this should be dated, and signed on the back by each of those who have helped in its preparation. The notes should be written up neatly in a permanent form: these also must be dated. Keep your map and note-books carefully, so that later on another group of scholars may make another survey and study any further changes.

Records of this kind have been made for all parts of Great Britain by the Land Utilization Survey working under Professor

Dudley Stamp of the London School of Economics. He gives a full account of the work in *The Land of Britain, its Use and Misuse*, a very useful book for the school library.The maps were made in 1930 and 1931 and the letters and the colourings were the same as on yours. Many of them have been published and you should, if possible, obtain one for your district so as to see what changes have taken place. Figs. 69 and 70 show two of the working maps.

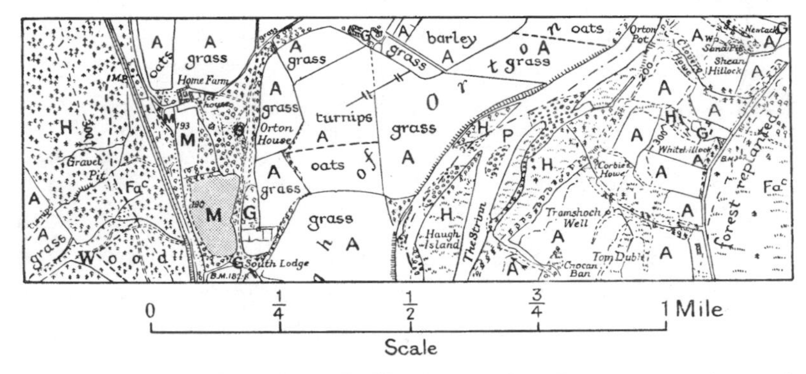

Fig. 70. Working sheet, Land Utilization Survey, Orton, Lower Speyside.

A, arable land; H, heath, moor, or rough grazing; M, grassland; Fa, high forest; Fa^c, coniferous.*

It may be possible to borrow old estate or parish maps for comparison with yours. Some of these go back to the days before all the present farms were made and show the large open fields divided into strips like big allotments on which the villagers used to grow their food before the enclosures were made. Usually there were three open fields, and in some villages each became a separate farm, so that the boundaries of the farm and of the open field are nearly the same. The Lord of the Manor used to have his own farm, shown on the old maps, and often that has survived, frequently keeping its

* F, forest; Fa, high forest; Fa^c, coniferous; Fa^d, deciduous; Fa^m, mixed; Fc, scrub.

old name: Manor Farm, Home Farm, etc. If you can make a tracing of one of these old maps it will add greatly to the interest of your collection.

During the course of the work you will often hear that one field is good wheat land, another is good barley land, a third is good potato land, and so on. You should record these, and find out the reason why from the farmer. There is always a reason for everything in nature, and one of the great purposes of agricultural science is to discover as far as possible what the reason may be. 'As far as possible'—for we can never get to the end, and however long and deeply we study even the simplest thing we can never know all about it. But we must keep on trying.

APPENDIX I

Mr H. V. GARNER'S MEMORANDUM ON PLOT EXPERIMENTS IN SCHOOL GARDENS

ROTHAMSTED EXPERIMENTAL STATION, HARPENDEN, HERTS

Choice of land. This is perhaps the most important point of all, for if the site is seriously defective, it is quite impossible to overcome this by subsequent manipulations, no matter how skilfully carried out. The first essential therefore is to examine the proposed site from the following points of view:

(1) Uniformity of soil (cultivation depth) and subsoil (to 9 inches below the base of the surface soil). The surface soil should all be alike in type and depth. If part of the area under examination is quite clearly different in texture from the rest, this is enough to rule out the site. For the subsoil more latitude is allowed, but if the subsoil varies from sand to clay, or contains local patches of gravel this also is a good reason for rejection. To get the necessary information a detailed soil survey of the area is required involving examination of about sixteen surface samples and eight subsoil samples taken so as to cover the whole plot uniformly.

(2) Uniformity of previous cropping and manuring. This is often a counsel of perfection, but sites having profound and drastic inequalities must be set aside. Thus if part of the land is old arable and the remainder has just been broken up from grass; or if part has been under old herbaceous border and the remainder under vegetables; or if one section only has been heavily manured with dung, compost, or lime within the last two years, the site will be unsuitable. The degree of latitude that can be allowed in respect of inequalities due to previous treatment is largely a matter of experience, but when in doubt it is always best to make another choice.

(3) Other inequalities. The following are to be avoided: (*a*) Areas overshadowed by trees, buildings, fences and so forth. If the site cannot be changed, a distance equal to the full height of trees and buildings and twice the

height of hedges should be allowed before the experiment begins. (*b*) Sites of old ditches, paths or trees now filled up, of old manure heaps or bonfires, 'made up' land in general. (*c*) Patches of persistent perennial weed.

AREA REQUIRED. The smallest area likely to be useful should accommodate:

16 plots each of 1/240 acre, total area 325 square yards.

or 25 plots each of 1/240 acre, total area 504 square yards.

These suit crops such as leeks, red beet, carrots, onions, white turnips and so forth where the individual plants are grown close together. Alternatively:

16 plots each of 1/160 acre, total area 484 square yards.

or 25 plots each of 1/160 acre, total area 756 square yards.

Suitable for crops grown in wider spacing as potatoes, sugar beet, swedes and the like.

The smaller the plots the higher is the standard of workmanship and manipulation required to attain a satisfactory degree of accuracy.

WORKING INSTRUCTIONS. Useful problems for investigation and the detailed instructions for setting up the experiment are, if desired, provided by the Rothamsted staff, and should be followed exactly. They will include: (*a*) plan of the lay-out of the experiment, giving the treatments to be applied and their position in the experimental area; (*b*) dimensions of plots and number of rows per plot; (*c*) weight of the manures to be applied to each plot; (*d*) any special instructions in regard to the application of manures.

CULTIVATIONS. In preparing the experimental site for the seed or plants uniformity in the nature and depth of the major cultivations should be secured. It would be wrong, for example, in breaking up new land to dig in the turf on part of the site and remove it from other parts. When the necessity for uniformity has been appreciated these matters should present no difficulty.

MARKING OUT THE PLOTS. Corners should be set out at right angles with an optical square, or by means of the chain as follows:

To set out a right angle *PAQ* at *A* using a chain or tape of 100 links, place link 32 at *A*, extend the chain so that link *O* (the handle nearest link 32) is at *P* in one arm of the required right angle. Place link 96 also at *P*. Take link 56 and pull it in the general direction *Q* till both arms of the chain are quite straight and tight. Link 56 then takes up position *Q* so that *PAQ* is a right angle. The arms of the angle may be extended to any required distance by sighting.

Measurements should be to within an inch or two, and should be carefully checked. The four main corners of the experimental area (not of the individual plots) should be *permanently* marked by posts, metal rods, or wires that *cannot* be moved by passers-by. The most convenient method is to use small wooden pegs 6 inches long to which stout wires 3 feet long are attached. The peg is buried at least 1 foot below ground-level and the wire stands above the surface to mark the corner of the plot. These corners should be located by measurements from surrounding fixed objects, gate-posts, railings, tree stumps, etc.

WEIGHING OUT AND APPLYING MANURES. This requires strict supervision. Make sure that each manure is correctly identified, and weighed out with sufficient accuracy, i.e. to within about 2%. Thus for a plot of 1/240 acre a dressing of 2 cwt. of fertilizer per acre would be approximately 15 oz. To weigh this to within an ounce would be too crude, a scale that moves for ½ oz. would be required, or alternatively, the units might be converted into grams and a rough chemical balance used. With farmyard manure at 16 tons per acre on plots of 1/160 acre the actual dressing of 224 lb. need be weighed only to within 7 lb. or so.

When one plot is scheduled to receive several kinds of fertilizer the appropriate amounts of each should be weighed out separately and then, if no instructions to the contrary are issued, they may be mixed together and transferred to a single bag for application. Sometimes all the plots in an experiment have the same quantity of a 'basal dressing' of one or more fertilizers, the object being to give a uniform non-experimental manuring to the whole area of the trial. In addition, each plot will receive a further dressing of some fertilizer under test. It is important to apply the basal dressings *plot by plot* in measured amount and not attempt to dress the whole experimental area from a single lot of fertilizer weighed out to the total amount required.

Before any manure is spread it is *absolutely essential* for a competent person to ascertain that the bag of manure is on its correct plot. A mistake here will completely ruin an experiment. Application can then proceed, the limits of each plot being marked by strings drawn tight between the corner pegs. Uniformity in application is important, this is a matter of careful and unhurried work. A common mistake is to give an undue proportion of the ration to the outside edge of the plot. A good procedure is to divide the dressing into two approximately equal parts. One of these is spread over the whole plots working lengthways and the remainder working crossways. The plot is thus covered twice. Any material left over should be cast over the whole plot.

SOWING THE SEED. The object should be to secure a full plant uniformly spaced in regard to distance between rows and distance between plants in the row. The ideal would be to have every plot alike in spacing and population. If the crop has to be thinned the above consideration should be borne in mind.

AFTER-MANAGEMENT. This is merely a matter of clean and uniform cultivation. If any plots are allowed to become much more weedy than others the results will be prejudiced.

GROWTH OBSERVATIONS. To derive the best from an experiment it should be closely observed at about monthly intervals starting at germination. The notes should be taken plot by plot on such items as: (1) germination, (2) colour, (3) growth; and any other noticeable feature. The observations are recorded by an arbitrary scale of marks from 1 to 10. Finally these numbers can be assembled and combined in various groups to bring out the effects of treatment and the validity of the results can be tested statistically. All marking of this kind should be done *without knowledge of the treatments*; bias and preconceived notions are thereby excluded. The marks themselves, when subsequently related to the treatments, will reveal any effects that may have been produced.

HARVESTING. The ideal procedure is to weigh the produce of every plot on the same day. This may not always be possible. Every experiment consists of several natural units, or groups of plots containing all the treatments under examination. These are called blocks or, in a Latin square, rows or columns. In harvesting therefore one (or more) complete blocks, rows or columns should be weighed on one day. The effects of any changes in weight that may happen between successive dates of harvesting, which would otherwise inflate the experimental error, can then be eliminated.

There is need of close supervision during harvesting operations to ensure that the produce of each plot is kept entirely separate. The crop must be completely secured, no potatoes, for example, should be left in the ground. The tops of root crops should be cut or wrung off and the roots should be weighed reasonably clean; the earth adhering to roots and potatoes can account for quite a high proportion of the total weight, and this proportion is far from constant from plot to plot. Hence the need for a certain amount of rubbing or rough scraping. The data required for each plot at harvest are usually as follows:

(1) Weight of clean roots (potatoes, parsnips, swedes, turnips, sugar beet, red beet, carrots, onions).

(2) Number of roots (for all crops except potatoes).

(3) Weight of tops (for all crops except main crop potatoes).

(4) Area of each plot harvested, or data from which the figure may be calculated.

Items (1) and (3) are clearly required; item (4) is necessary in order that weights may be calculated on an acre basis. This information is very often overlooked and much needless delay and correspondence are thereby caused. Since almost all crops likely to be experimented with are grown in equally spaced rows the best way to specify plot size is *mean* row width × number of rows harvested per plot × length of row harvested per plot. The mean row width is ascertained by counting the total number of rows across the experimental plot and dividing this into the width of the plot. Item (2) is to enable the gross effect of variation in plant population to be allowed for. For most crops there is no difficulty in counting every root weighed, but carrots often give such a high plant population that even to count the whole produce of a rod of ground is somewhat laborious. In this event an estimate of the plant population per plot may be obtained by counting the roots in each of a number of 'sampling units' consisting of one yard of row. These units are located at random within the plot boundaries, and altogether they should account for about one-tenth of the crop or about 200 roots, whichever is the smaller. Given the length of row and number of rows per plot, directions for sampling may be obtained from Rothamsted.

SPECIAL CROPS. Most of the crops grown will be root crops, but cabbages, lettuces, leeks, brussels sprouts and tomatoes, may also come in. Here, in addition to the above data, there will be some measurements specifying quality, for example:

Cabbages: (1) weight of total cabbage as cut; (2) weight of hearts.
Lettuces: (1) weight of hearted lettuces; (2) weight of open lettuces.
Leeks: (1) total weight; (2) weight of trimmed stems.

Sprouts: (1) weight of hard sprouts in successive pickings; (2) weight of blowers.

Tomatoes: (1) weight of ripe fruit in successive pickings; (2) weight of green tomatoes at end of experiment.

Any other data of a similar kind may be recorded plot by plot and sent in for statistical analysis. For example: with carrots the number of split roots may be counted and weighed; with onions the number and weight of bull necks; with potatoes the weight of the total crop that passes through a riddle (whose size must be recorded); with cabbage or lettuce the numbers of first-class marketable plants at one or more pickings. Counts of diseased tubers may also be made.

RECORDING OF RESULTS. In the field the best procedure is to number the plots serially on a plan of the experiment and record all observations under the appropriate plot number. Normally the weights of produce will always be 'produce + container', the weight of the container being ascertained and recorded as well as the total weight. The necessary subtractions may be left till later. The main point is to record systematically the original figures as weighed in the field. Bold legible writing that will survive the effects of rain and mud is essential.

The weights should be added up for each plot as it is finished. A particularly high or low plot total will be observed at once and can be checked or confirmed—in this case reference to the plot treatment will be of assistance.

WORKING OUT THE RESULTS. Experimenters should attempt to extract all the information the data contain. Rothamsted is prepared to help where the experiments have been carried out in co-operation with the Station; it then needs the primary data as taken in the field under plot numbers, accompanied by a numbered treatment plan, or a plan of the experiment in which the weights and treatments are entered in their appropriate squares. If this method is adopted it cannot be too definitely stated that the transcription from field note-book to plan must be correct.

We will assume that a set of say twenty-five yield figures each, representing the produce of a single plot, has been provided. These yields differ between themselves and exhibit variations which may be assigned to different causes:

(1) Variation in yield due to the experimental treatments applied. These are the fertilizer effects that are being studied and are the essence of the

experiment. In so far as they exist they are contained in the comparison of the total yields of groups of plots sorted out according to their treatments. This is the usual common-sense method of handling the results.

(2) Variation due entirely to accidental causes and not connected with the effectiveness of the manures. Some of the main causes of this variation are: (*a*) differences in the quality of the soil from place to place within the experimental area; (*b*) differences in manipulation of the plots in respect of cultivation, measurement of areas, weighing and application of manures, sowing of seed, after-care and harvesting; (*c*) differences in extent of damage by pests and diseases.

The items in group (1), i.e. those particularly sought out for measurement and study always carry a component of uncertainty due to the combined effect of the items in (2). This component may be isolated statistically, and the appropriate figure is usually called the Standard Error. A low standard error, other things being equal, means a precise experiment and vice versa. Clearly, therefore, the manipulation of the experiment should be directed towards reducing the disturbing effects of the items in category (2) to a minimum. Some of the methods adopted for this purpose have already been outlined.

As an illustration of the effect of the magnitude of the standard error on the value of the experiment the following may be given. Suppose two experiments on potatoes each gave the same average yield (taken over all plots) of 8 tons per acre, and one has a standard error per plot of 8 per cent of the mean yield, the other 30 per cent. (The first represents the degree of accuracy attained by ordinary good workmanship, the second would be regarded as a very crude trial indeed, where high accidental variations have occurred.) If each of the experimental treatments rests on four repetitions then in the first experiment an effect of manures would be judged 'significant' (i.e. due to the action of the experimental treatments and not to the accidental variations previously mentioned) if it attains the value of 0·96 ton per acre; while in the second experiment a difference of 3·60 tons per acre would be required in order to be accepted with equal certainty. Now differences of the order of 1 ton or more are frequent in experiments on potatoes, but differences as big as 3·6 tons are rare. Consequently, only quite exceptional treatments would show statistically significant effects in so inaccurate an experiment as the second, and the time and trouble spent in carrying it out would be largely wasted. The relative values of similar experiments vary inversely as the square of the standard errors.

AGRICULTURAL NOTES AND GENERAL INFORMATION. The numerical results taken alone do not tell the whole story, indeed without subsidiary information they lose much of their value. For every experiment the following data are required: (1) crop; (2) variety of seed; (3) date of applying artificial and organic manures; (4) method of incorporating manures with the soil, i.e. raked in, dug in, put down the ridges, etc.; (5) date of sowing seed; (6) date of harvesting; (7) previous crop; (8) general notes of nature of growing season and any pests, diseases, storms, etc., that may have affected the experiment as a whole or certain plots in it.

(5) Where the material sampled is moist as in the case of dung or sewage sludges the sample should be packed in a tin to prevent water-loss on the journey to the laboratory. This does not apply to soils.

ASSISTANCE AND ADVICE. Mr H. V. Garner, Rothamsted Experimental Station, Harpenden, Herts, is always willing to advise in regard to plot experiments. Before writing to him, however, teachers should read the chapter on Field Experiments in *A Student's Book on Soils and Manures*, E. John Russell (Cambridge University Press), where the underlying principles are set out. A fuller exposition is given in Technical Communication no. 35 (Technique of Field Experiments) Commonwealth Bureau of Soil Science, Harpenden, which also should be read.

APPENDIX II

BOOKS AND QUESTIONS

The teacher is advised to procure, both for his own information and in order to read passages to the scholars:

Gilbert White, *Natural History of Selborne.*

Charles Darwin, *Earthworms and Vegetable Mould* (Murray).

A. D. Hall, *The Soil* (Murray).

E. J. Russell, *A Student's Book on Soils and Manures* (Cambridge University Press); *Soil Conditions and Plant Growth* (Longmans).

L. Dudley Stamp, *The Land of Britain, its Use and Misuse* (Longmans); *Britain's Structure and Scenery* (Collins).

Mr Hugh Richardson's list of questions, worked through at Bootham School, York. They have proved useful to other teachers also.

1. COLLECT SAMPLES of the different soils in your neighbourhood—garden soil, soil from a ploughed field, from a mole hill in a pasture field, leaf mould from a wood, etc. Collect also samples of the subsoils, sand, gravel, clay, peat.

2. Supplement your collection by purchasing from a gardener's shop some mixed potting soil and also the separate ingredients used to form such a mixture—silver sand, leaf mould, peat.

3. How many different sorts of peat can you get samples of? Peat mould, peat moss litter, sphagnum moss, turf for burning, dry moor peat?

4. Find for what different purposes sand is in use, such as mortar making, iron founding, scouring, bird cages, and obtain samples of each kind.

ANALYSIS OF GARDEN SOIL. About a handful of soil will be required by each pupil.

5. Describe the appearance of the soil. Is it fine or in lumps? Does it seem damp or dry? Can you see the separate particles of mineral matter? How large are these? Is there any evidence of vegetable matter in the soil?

6. Put some of the soil in an evaporating basin and over this place a dry filtering funnel. Warm the basin gently. Is any moisture given off?

7. Dry some of the soil at a temperature not greater than that of boiling water, for example, by spreading it out on a biscuit-tin lid, and laying this on

a radiator. How have the appearance and properties of the soil been changed by drying?

8. Crumble some of the dried soil as finely as you can with your fingers. Then sift it through a sheet of clean wire gauze. What fraction of the soil is fine enough to go through the gauze? Describe the portion which will not pass through the gauze. Count the number of wires per linear inch in the gauze.

9. Mix some of the soil with water in a flask. Let it stand. How long does it take before the water becomes quite clear again?

10. Mix some more soil with water. Let it settle for 30 seconds only. Pour off the muddy water into a tall glass cylinder. Add more water to the remaining soil, and pour off a second portion of muddy water, adding it to the first, and so on until all the fine mud is removed from the soil. Allow this muddy water ample time to settle.

11. When the fine mud has settled, pour off the bulk of the water; stir up the mud with the rest of the water; transfer it to an evaporating basin, and evaporate to dryness.

12. Does this dried mud consist of very tiny grains of sand or of some material different from sand? Can you find out with a microscope?

13. If the mud consists of real clay and not of sand it should be possible to burn it into brick. Moisten the dried mud again. Roll it if you can into a round clay marble. Leave this to dry slowly for a day. Then bake it either in a chemical laboratory furnace or in an ordinary fire.

14. Return to the soil used in Question 10, from which only the fine mud has been washed away. Pour more water on to it, shake it well, and pour off all the suspended matter without allowing it more than 5 seconds to settle. Repeat the process. Collect and dry the poured off material as before. What is the material this time, sand or clay?

15. Wash the remaining portion of the soil in Question 14 clean from all matter which does not settle promptly. Are there any pebbles left? If so, how large are they, and of what kind of stone?

16. Take a fresh sample of the soil. Mix it with *distilled water* in a flask. Boil the mixture. Allow it to settle. Filter. Divide the filtrate into two portions. Evaporate both, the larger portion in an evaporating basin over wire gauze, the smaller portion in a watch glass heated by steam. Is any residue left after heating to dryness?

17. Take a fresh sample of soil. Spread it on a clean sand bath and heat strongly with a Bunsen flame. Does any portion of the soil burn? Is there any change in its appearance after heating?

18. To a fresh sample of soil add some hydrochloric acid. Is there any effervescence? If so, what conclusions do you draw?

19. Make a solution of soil in distilled water, and filter as before. Is this solution acid, alkaline or neutral? Are you quite certain of your result? Did you test the distilled water with litmus paper? And are you sure that your litmus does not contain excess of free acid or free alkali?

PEAT

20. Examine different varieties of peat collected (see Question 2) and describe the appearance of each.

21. Burn a fragment of each kind of peat on wire gauze. What do you notice?

22. Boil some peat with distilled water and filter the solution. What colour is it? Can you tell whether it is acid, neutral or alkaline? Evaporate some of the solution to dryness.

OUT-OF-DOORS

23. Describe the appearance of the soil in the flower beds: (*a*) during hard frost, (*b*) in the thaw which follows a hard frost, (*c*) after an April shower, (*d*) in drought at the end of summer, (*e*) in damp October weather when the leaves are beginning to fall.

24. Is the soil equally friable at different times of the year?

25. In what way do dead leaves get carried into the soil?

26. Can you find the worm holes in (*a*) a garden lawn, (*b*) a garden path?

27. Take a flower bed or grass plot of small but known area (say 3 yards by 2 yards) and a watering can of known capacity (say 3 gallons). Find how much water must be added to the soil before some of the water will remain on the surface. What has been the capacity of the soil in gallons per square yard?

28. Take two thermometers. Lay one on the soil, the other with its bulb 3 inches deep in the soil. Compare their temperatures at morning, noon and night.

29. Find from the 25-inch Ordnance Survey map the reference numbers of the fields near your school. Make a list of the fields, showing for what crop or purpose each field is being used.

INDEX

INDEX

Printed in the United States
By Bookmasters